HANNAH STOWE has a degree in marine biology and is a qualified and experienced sailor. She currently lives in Germany, painting, writing and sailing her own boat.

MOVE LIKE WATER

A Story of the Sea and its Creatures

HANNAH STOWE

GRANTA

Granta Publications, 12 Addison Avenue, London W11 4QR
First published in Great Britain by Granta Books, 2023
This paperback edition published by Granta Books, 2024

A CIP catalogue record for this book
is available from the British Library.

1 3 5 7 9 10 8 6 4 2

ISBN 978 1 78378 860 6
eISBN 978 1 78378 861 3

Typeset in Caslon by M Rules

Printed and bound by CPI Group (UK) Ltd, Croydon, CR0 4YY

www.granta.com

To Jackie Morris, Nicola Davies, and Jessica Woollard.
My three graces, three furies, three witches.

Contents

MOVE
LIKE WATER

I

Fire Crow

There was never a time when I did not know the sea. As I lay in my cradle at my mother's feet, day after day, the salt wind blew around our home. It mingled with the honeysuckle that curled around her garden studio, sweet-scented and dappling light as she coaxed gentle worlds to paper with paint. The small strong oak trees my father had planted when I was born bent and twisted to that wind, framing my world. A hushed roar, water on sand and stone as the tides ebbed and flowed, both rhythm and rhyme. At the start, it was only a lullaby.

Throughout my childhood the weather was never far away. At night, I would nestle in my bed, tucked under the eaves in the attic of our cottage, snug next to the chimney breast, fire-warmed, as storms shook the slates from the roof, a ginger cat purring beside me. As I lay awake, I would watch for the beam, the beacon, of Strumble Head Lighthouse as it swept through the night, my companion in the ink hours. In the morning I would scrape the salt spray from the windows, running my finger through its grey-white slick, a quick, sharp taste on my tongue.

That cottage by the sea was a harbour of sorts, a place I always felt safe. It was ramshackle in every way – rendered wall crumbling, paint flaking, wallpaper peeling, furniture clawed by cats and carpets chewed by dogs. The garden was a blend between overgrown and functional, my mother's wild resistance to the manicured lawns on which she was raised. There was a herb garden and vegetable beds, but they weren't always tended. Brambles and alexanders grew in the hedges, and you had to watch for nettles. A great ash tree stood like a guardian at the foot of the path, its leaves lush and verdant. At night, I would often stand under this tree and look up at the night sky with my mother, locating the constellation of Orion, the great hunter with his sword, bow and belt. There was no reason for Orion in particular, save that those were the stars I was drawn to first. I would stare up at them, fascinated by the celestial light. A concrete path chalked with hopscotch, crossed and noughted, snaked to the door. Ants would track across it, industrious and purposeful, before returning to their nest opposite the rose bush. Once, when he was visiting us, my grandfather, a contradiction of a man – a naturalist when out walking the Cotswold hills where he lived, naming every bird that winged its way past, every tree, from leaf and bark, as if they were his own family, and yet an enemy to the rhythmic chaos of nature in his own garden – threatened to pour pest killer into the ants' mound. The day he left, I reached a small fist up to the kitchen shelf and took a bag of sugar, the bleached granules sparkling in the sun as I poured out a veritable feast in rebellion.

Chickens pecked in the yard, sweet peas climbed canes and dog roses bloomed. An orange creel buoy hung above the oak front door, the window frames were decayed from salt, and the panes rattled. The weather vane sitting atop the roof was askew, 'north' pointing more to north-east than the pole. Inside the house, mountains of books formed dusty labyrinths, the washing-up was usually in various stages of undone, and my ginger cat slept on the mantle above the hearth, looking like Dürer's hare. Finding a place on the sofa required delicate negotiation with a collie dog or lurcher. The only source of heat was the fire in the hearth that burned year round. I would sit in front of those flames, a sheepskin rug on the cold slate floor, with pens in my hands, brushes, paper, my raging moods that would often flare turned to paint and ink; even then, I found expression through written word and paint far easier than making myself understood aloud.

There was a current inside me. At times, it swept along straight and true, serene on the surface, but determinedly fast flowing. At others, the winds of life would turn against the tide, steep over falls would whip up in seconds, and I would rage, tempestuous. And sometimes. Sometimes the water flowed to a deeper place. A place where I could still see the light streaking through from the surface, and yet I felt compelled to stay quietly in the dark. This was, this is, my nature. Days and years added length to my limbs, strength to my muscles, and, all the while, the salt song grew louder. At night, as the lighthouse beam rotated, roaming both land and sea, I began to wonder how far the light stretched, what was out there in the reaches of that dark. I started to climb the trees, high, high, higher every time, until the branches grew thin and supple, bending under my weight. Higher still, pushing the boundaries of my garden, my range, up into the air, swaying as if in the rig of a ship, until at last there was no canopy left to claim. From the branches at the top, I could see over the roof of the cottage, all the way to the sea. It wasn't long before I started to climb out of the attic-room windows, on to the roof slates themselves for a better view. Once every corner of the cottage, inside and out, had been explored, I started to wander further, venturing past the hedges frothed with blackthorn blossom that bordered my home.

Red Kite,
Sky Blue,
Gorse Yellow.

These were the primaries of my landscape.

There were two paths to the sea. One ran along the borders of pastoral land, tamed and grazed by shaggy tack sheep, wisps of their wool flying on the barbs in the wire. The other had a more rogue nature, a stone-wall-lined lane, thick with ferns and navelwort, lush dark green against grey, falling away to hawthorn, with its tiny red lanterns that would ripen and burn hope into the landscape all winter long. At the top of the lane, through the gate, the landscape opened out, gorse and bramble ready to snatch at skin. Wild ponies roamed, spirited and unafraid to give chase, while birds of prey wheeled overhead. The south-westerlies blew strong, the entire Irish Sea laid out in front of you, vast and shimmering, light, dark, stormy, still. From here, you could descend further, until the path became either fox track or badger trail, snaking through bracken, becoming stone as you neared the edge of the cliff to scramble down to the sea. As a teenager, I would stand at the brow of the hill, eyes closed, feet rooted firmly, grounding. I would breathe in the land, old stone and wild water. Breathe it in until my chest hurt, full to bursting. With a whoop, I would open my eyes, take off at a sprint, gaining momentum with every step as I careened towards the clifftop. It always felt as if, with enough courage, I could launch myself over the sill, the thin yet firm divide in matter, where land gave way to air over water, and wing my way to flight with the fulmars. Every time, the pull of the earth would stop me, and keep me for its own. That path seemed to speak to the wildness that sometimes overtook me during those years. But when I first started exploring, I chose the simpler, safer route.

That way took me over the stiles, past golden stubs where the hay had been cut and baled. Fields became a path, the path an orderly set of steps. A stone bridge crossed a stream, fresh and flowing towards the salty expanse. I came to know the coast here in all its many moods, but the constants were the long reach of St David's Head, stretching west, a tidal race rushing around the point. On the horizon, the rocks and islets known as the Bishops and Clerks: North Bishop lying like a sleeping giant in the ocean, South Bishop with its lighthouse and accompanying keeper's cottage. Ramsey Island was closer, only a few miles away, laid out towards the south-west. On clear days, you could see all the way out to the Smalls Lighthouse, a little needle on the horizon.

There was a gently washed riddle in the ridges of the sand, the day balmy, the water shining, cool, inviting. I tripped over the sands, my legs tired from this early odyssey towards the sea, stumbling steps that became a run to the tideline. My toes touched the water first, droplets flying in front of me, jewels in the air. The cold bit, but I didn't stop yet. Not until the water had passed my knees, pooling up to my waist, edging deeper, until that moment of grace where the water swept me up, holding my weight. The surface a shining skin, I was so convinced I could peel it back. But of course my hand went straight through, one second in air, the next in water. The sun warmed my shoulders, but the shivers were starting to set in. With a settling breath, I dived headlong into the swell, exhilarated by the surge of speed. Eyes open underwater, the world a blur in soft focus, a subtle blue. The salt stung my eyes. Light streaked through the water, golden kelp streamed, tangling with my hair. Earth tones in the ocean. I was young enough that I didn't think – I knew, and I knew water. In time, I would find a solace and a synergy in the sea, but then it was as natural and as essential as breathing. I had learned to swim at the same time as I learned to walk, frog legs kicking behind dog-eared floats in the community pool and in the sea. We all learned, all the children of this coastal community, surrounded by the sea on three sides. We learned about rip tides, what to do in a strong undertow. Learning the cycles of the tide came with learning to read a wristwatch, if not before. There were greater ramifications to getting cut off by a rising tide than to knowing when a quarter past four was. It wasn't until I was seventeen that I met someone my age who couldn't swim.

Spring would sweep the coast path with bluebells, blooms thronged with thrift and squill that would push themselves through the ground with the energy of renewal. As the days grew longer, the sunlight stretching into the evenings, my brother and I would scramble the cliffs on to the rocky shore to line fish for the mackerel that would arrive on our coast as the water began to lose the chill of winter. The line between land and sea seemed to vanish as we cast our hooks through the air and out into the water. We would pick sorrel and samphire – sweet, salty, waxy, fresh, all at once – and wrap our fishy parcels in foil with butter and lemon to cook over a fire. Swims in spring were bitter and biting, the sun not yet strong enough to give warmth.

Often I would walk the tideline with my mother, shivering

and wrapped in layers of wool, cold toes tucked into warm boots. We were searching for sea glass, razor clams and mermaid's purses, the discarded egg cases of elasmobranchs. Once I found a hag stone, a pebble with a perfect circle worn through the centre of it. My mother said if I looked through the hole I would see the truth of the world. The rocky shore was a treasure trove, the blanket of the sea pulled back to reveal the tide pools. With my ear close to the rock, I would listen to the hissing pop of barnacles as they closed their shells, sealing inside a tiny sea, so they didn't dry out in the sun while waiting for the tide to return. The pools themselves were littered with jewel-like anemones, red and glistening, fat and ready to burst, or tentacles waving in the water. I would reach out a finger, feeling the gentlest of tugs as the anemone tried and failed to pull my finger to its mouth, all the while firing stinging cells, nematocysts, into the skin – a tiny sea monster trying to devour me whole. My finger was too strong and my skin too thick and I was free from the anemone's clutches. I found pools that were starscapes, their constellations formed from echinoderms, cushion stars. In others, sand-coloured blennies would flit from my shadow when I first peered down, becoming emboldened if I sat still for long enough and watched. Top shells crowded, limpets clung, shore crabs skittered, each pool its own little world, islands in the inverse, water from land rather than land from water, before the tide returned to connect them all once again.

Early summer brought lighter mornings, the sun banishing the lighthouse beam and breaking through the salty glass of my window in the small hours. Every morning I would ride my bike to the beach, collecting pastries at the farm shop on the way to eat in the sunshine after my swim. The sea was still cool, but it no longer stole my breath when I waded in.

A seasonal bounty of plankton would start to bloom after the winter's rest, marine wanderers in the water, creating swathes of speckled clouds, like the dust that was ever present in our cottage. Jellyfish followed, feeding on the plankton. There were moon jellies, *Aurelia aurita*, transparent, marked by four pink rings on the bell. Their sting was harmless, but you needed to be more wary of the lion's mane, *Cyanea capillata*. A blooming, burned-orange bell, textured on the underside that billowed in the water, long white trailing tentacles that could stretch for metres, longer than a blue whale in the colder arctic waters, the sting a sharp burn which has left thin scars on my skin. There were comb jellies, tinged pink, gooseberry-like in morphology, flicking out ethereal phosphorescence with their cilia, blue, red, green, yellow, glowing. Spider crabs swarmed into caves and sheltered spots in order to shed their shell, a moment of vulnerability as they left their hard casing in order to grow. I would snorkel over their teaming masses, a tangle of limbs waving. The discarded shells, composed from chitin, calcium carbonate and protein, doubled the apparent numbers, but if you looked closely you could tell which of the spiney orange exoskeletons were inhabited, a telltale stillness giving away the empties.

Before long, seabirds would begin to return to our shores; Manx shearwater, the most elegant of wing, making a long gliding journey north from South America, up the eastern seaboard, following the Gulf Stream, to take up residence in burrow nests on Skomer, Ramsey and Skokholm. Puffins also made underground shelters on Skomer and the North Bishop, while gannets in their tens of thousand would paint the island of Grassholm white with wing, a cackling cacophony of birds that set the air to sting with the ammonia of their guano. Guillemots and razorbills, with their small bodies and staccato wing beat, found security in the most precarious of roosts, perching on slithers of ledges that laddered the cliffs, too small for greater black-back gulls to land. Kittiwakes built their small nests around the mouths of caves. The fulmar petrel, a true seabird, with stiff-winged, gliding flight, designed to travel effortlessly over the ocean, would choose an exposed cliff face for a nesting site. This was the only time they would come ashore, using the updraught as the sea air hits the cliff to come into land, taking to the wing again as soon as their chicks fledged.

Seals arrived in their hundreds, well-fed and ready to pup on the island beaches and in the secluded bays of the mainland. The seals were Atlantic grey in name, but their dense, mottled waterproof fur covered a spectrum from shining silver or earthy brown to deepest black. The dorsals of common dolphin broke the surface in St Brides Bay, hourglass patterning down their flanks, looking so exuberant and carefree. They would calve here too. Sometimes, if we were lucky, we would see them mere days after birth, still tiny, their skin etched with foetal folds from fitting into their mother's womb. Occasionally, you would see the large fin of a Risso's dolphin, the stories of their ecology carved on their skin in white scars against grey, telling of encounters with the beaks of the oceanic squid on which they feed. We would see the ancient form of the *Mola mola*, its disk-like body laterally flattened, waving its flopping fin as though hailing to the sky that gives it its names: 'sunfish' in English, 'moonfish' in Italian. Once, from the coastal path at St David's Head, I saw the gaping mouth of a basking shark, wide open, ridged and cartilaginous. Harbour porpoise would surface with puffs of air, feeding in the flow of the tide.

The abundance of summer was everywhere, infectious. The sleep of spring had been shaken from the quiet coastal villages, as suddenly the population doubled, tripled, quadrupled. Barbecues were lit in the gardens of cottages whose hearths had not seen a fire all winter. Different voices carried through the air, less lilting, *a*'s stretching to *arghs*. The cars were shinier, less rusted by the salty breeze, and queues in the shops longer. The beaches that had been my quiet queendoms all winter long filled with people; caravan sites became cities, with suburbs of colourful tents. There is always a tension palpable in the otherwise carefree air of summer, an 'us and them'. I didn't pick up on this as a child, perhaps because my brother and I were the first of our family to be born here, and our parents were English, without the generational branches of our family tree stretching into Pembrokeshire soil. I still felt as if I had absorbed the shape of the land, the water, into my bones, that this place had given me my foundation, but it wasn't mine and mine alone, to own and hoard. I loved to share the seascape, to see the delight of others as they got to submerge themselves in what was my everyday. It wasn't until much later that I realised the privilege of my upbringing. I grew up with self-employed parents who made their own work, so I did not yet understand the stress of seasonal businesses, where hectic opportunities would dry up come autumn. We owned our little cottage, so I knew nothing of the difficulty of finding a year-round rental when holiday lets were far more profitable, which was the lament of people who had grown up here on looking through the windows of empty houses, second homes – the prices were unreachable on a local wage. Later, I came to understand the frustration more deeply, but as a child what I saw, what I liked, was that there was a little more colour around, sometimes different languages, different faces.

If I wanted to escape the crowds, I could find solitude in swims at dawn, or head to beaches that were hard to find, easier to know. One of my favourites was marked on a map as Aber Llong, but we all called it Tug Beach. The path is subtle, it slips over the cliff in a way that puzzles those who don't know it's there. Down a scree slope that promises aching legs on the ascent, and on to a large, sloping tongue of stone, dark black, brown, and rippled from the touch of the sea. The boulders here are massive, and we would go above and below them, crawling through the gaps, our own world of tunnels. They get hot in the sun, burning the soles of your feet, but perfect to haul yourself out on and sun bake after a swim, the salt crystallising on your skin. Ten years before I was born, three tug boats, outward bound from Liverpool, had run aground and were wrecked here, washing up on this beautifully jagged stretch of coast. There isn't much left of them, just the engine blocks, really, rusting into the sea year after year, a reminder of the savagery of the Pembrokeshire coast, which has many such wrecks. To me, they were great dying beasts of cylinder and piston, the workings of their previous lives a mystery. A friend remembers being called down by her father to the shore to salvage the tin cans that had washed up. The sea had stolen the labels from them, and for weeks her dinner was a lottery – stew or peaches, peas or custard.

You would feel summer start to slip away in the air as the blackberries began to bud and ripen in the hedgerows, the dark creeping into the ends of the day. Tents were packed up, the crowds thinned, and some houses closed their doors and remained shuttered up for the rest of the year. There was a transience to the population. A lot of the jobs here were temporary, winter work scarce, the hours short and the pay bad. Tanned surf instructors would head off to the southern hemisphere in search of an endless summer and consistent swell, climbers who had scaled the sea cliffs would leave for snowy mountain peaks, and bar and restaurant staff went back to their university cities or to the Continent.

Most of the seabirds had already left. Every evening during

the summer, the Manx shearwaters would stream in monochrome, black backs, white underbellies, gliding over the surface of the water with innate grace, returning to hungry chicks in their burrows after a day at sea. But by the end of August, this daily avian spectacle would stop. Their chicks, either through boldness or hunger, would emerge into the light for the first time, and, by some unknown inheritance, ride the wind to South America to overwinter. We would see guillemot chicks making their own improvised departure. Their major flight muscles, the supracoracoideus and pectoralis major, were not yet fully developed, and their wings were too small to propel themselves through the air, so they would fledge before they could fly, jumplings, leaping from their ledges to descend on to the water for the first time. From here, they would swim out to raft on the sea, fed by their fathers, who dived for sand eels and sprat until they could dive themselves. Razorbills and puffins too would take to the sea, far less easy to tell apart once they lost their breeding plumage and the brightness of their beaks had shed to dull. Gannet chicks, black-feathered to distinguish the juveniles from the parents' white plumage, too fat to fly, would start their migration away from the breeding colony in the water, swimming before they could take to wing. The young kittiwakes would find their nests being dismantled around them, a less than subtle sign from their parents when the time came to head offshore.

From late summer onwards, into the autumn, the seals would begin to pup. Pregnant females would haul out on the beaches, above the high-tide mark. We would watch for hours, carefully and quietly perched on the clifftops, peering through binoculars, as the pups slipped into the world, wriggling, white fur stained yellow with amniotic fluid. Born on land, the pups would not be able to take to the sea safely for weeks. As they latched to feed, gulls, forever bold, would pick through the afterbirth for the prize of nutrients. As the seal mothers birthed and fed, a single bull would patrol, claiming a cove, a cave, or a section of a beach for his own. For weeks, he would not let another male come close, guarding his territory without even breaking to feed himself. His claim for this protection was the right to mate with those females when they returned to the water again.

Birthing, weaning and breeding was a feat of bodily endurance for the female seals. After birthing, they would barely leave the side of their pup for weeks. For sixteen days or so, the mother would undergo a fast, losing her condition drastically, while the pup would become drunk on rich, 60-per-cent-fat milk. On the rare occasion that she gave birth to twins, it was likely that the mother would simply not feed one of them, rather than have two pups slowly starve as her milk supply dwindled. Beyond protecting the area where his female seals were nurturing their young, the male would have nothing to do with the rearing of the pups at all; his sole aim was to impregnate the females again, an interaction that would result in embryonic diapause (where the egg is fertilised but implantation into the uterus is delayed). This afforded the females time to recover from the birth and rebuild their physical strength for the next pregnancy, and ensured that their next pups would be born at the most favourable time of year. The duration of the diapause was typically three months, the gestation period nine, so, all being well, the females would haul themselves out on the very same beach the next year, birthing, fasting, feeding a pup, mating.

It is a relentless cycle, with the females experiencing postpartum oestrus, the ability to get pregnant directly after birth. I pondered why they have this extreme physical stress without rest, and the answer I found was that the necessity for this oestrus is largely due to the way in which these marine mammals move. Although they do not migrate as such, the grey seals disperse widely around the coast of Britain and through the Atlantic during the winter months, their movement patterns dictated by foraging. They go where the fish go. In the summer months, when the colony comes together, clustered in tight groups to pup, it is a highly advantageous time for breeding. The females find a competitive mate, a male who has won and successfully defended his harem; and the males gain access to the highest possible number of reproductively viable females. Afterwards, the adult seals, both male and female, will depart, and the newly weaned pups, freshly moulted to silver, will be alone to face the autumn swells and to learn how to forage. In good years, the weather stays balmy through September and into October. In the bad, the storms set in early and the mortality rate among the pups is high. In the absence of heavy predation on our shores, the sea decides the seals' fate.

Nature was nature, harmonious in its disharmony. Gulls would try and pull the auks from their ledges, or drown them, holding them under the surface of the water until the life was thrashed from them. Ravens, clever thief birds, stole eggs from nesting ledges and flew off with them between their beaks – bright flashes of turquoise eggshell taken from the guillemots, cream speckled with brown from the razorbills. More eggs than they could possibly eat, some to be buried and dug up later as sustenance through the winter. Most auk pairs would lay only one egg in a breeding season, so if the egg perished the pair lost their chance to pass on their genes, to raise a chick that year. Peregrines and buzzards snatched smaller birds in flight, making steep dives to claim their prey between sharp talons. I once saw a peregrine sitting in a nest of kittiwake chicks, the sickle-winged raptor like a knife blade amongst the fluffy gulls. Fulmar chicks were the exception to this pattern of early death. Once the egg had been incubated and hatched, a period of around fifty days, it was not uncommon to see a fluffy chick alone on the nest while both the parents spent the day foraging at sea. The secret to their survival was that if approached and threatened, the baby fulmars had a delightfully visceral reaction. They would projectile vomit on to their predator a sticky, foul oil, which would stick on contact with feathers and cause them to rot. Predatory birds knew to keep their distance from the fulmars' clifftop perches.

There is something else at play, however. As I grew up, I became increasingly aware of the impact of humans on the seascape and wildlife, and even the seasons. The natural cycles of weather are being disrupted, the play between predator and prey altered, competition between and within species is being affected. I could see it happening just eight miles from the Pembrokeshire coast.

Grassholm Island is a stronghold of *Morus bassana*, the northern gannet. In each breeding season, over 80,000 birds crowd together at the south-west edge of the rock, where the prevailing winds assist the birds with take-off. The gannets themselves are large – the largest seabird in the northern hemisphere – with a wingspan stretching to two metres. Their plumage is white, their sharp, streamlined wings tipped with black feathers. Their heads, gently crowned in pale yellow, are the only soft thing about the birds, everything else angles and points. Their eyes are a piercing, icy blue. It is thrilling to see the gannet circle and spiral, ever watching the surface of the water. In a moment, the great bird dives, plummeting headlong towards the skin of the sea. Sharp spear beak breaks the surface for the arrow bird to enter, wings dislocated in a split second as it shoots through the water. If a fish is caught, it is swallowed whole before the bird sits on the surface, allowing a moment for its prey to slide down its gullet before taking to the skies once again. White, black, yellow, flash of ice from the eye. But these are no longer the only colours to be seen on the island.

Now there is orange and turquoise from stringy plastic filament nets and fishing gear. The gannets used to build their nests using seaweed, but now, as they collect material, the pervasive plastic has crept in. It is estimated that there are around twenty tons of plastic on this isolated island. Gannet chicks hatch as downy puffballs, light feathers drifting on the air as they shed to their dark brown-black juvenile plumage. As they grow, they can become tangled in the net, the twine. It catches around beaks, snaring them shut and starving the birds. It tangles around feet, breaking their tiny bones and cutting off circulation, grounding them. Parents feed and feed their chicks, who simply cannot fledge, physically tied to the island by human carelessness. Everything thrown into the sea goes somewhere.

Seals too, in their curiosity and playfulness, can become tangled in lone floating ropes, ghost gear that has been abandoned by fishermen. Some will simply never surface again. I have seen young seals with twine caught around their necks. Grey seals continue to grow after maturity and well into their adult lives. As they grow, the twine cuts deeper and deeper into the blubber every year, causing red welts, fur rubbed away, until, if the animal does not manage to dislodge it, they will be strangled. Another seal I saw had a discarded frisbee ringed around its head like a grotesque halo. The entirety of the shelf waters around Britain are considered potential habitat for harbour porpoise, *Phocoena phocoena*. Both porpoise and seals alike are in danger of becoming by-catch – the unwanted fish, seabirds or marine mammals caught during commercial fishing, with no financial value. Unwanted. To be discarded. Gill nets, a wall of filament that sits in the water column, are the nylon nightmares of both cetaceans and pinnipeds. If a seal, porpoise, dolphin or whale gets tangled in one, they will suffocate, unable to reach the surface for air. Whale and Dolphin Conservation, the WDC, reports that in the UK alone at least 1,000 porpoise and 250 common dolphin are the victims of by-catch every single year. That number is ambiguous and the reality is very likely to be much higher, as by-catch can be and is discarded at sea, unreported and unrecorded.

Before I was twenty, I saw for myself the decline in the numbers of kittiwakes. There is a nesting site for these small onomatopoeic gulls around the mouth of a cave called Ogof Cantwr on Ramsey Island. The cave is a cathedral carved out of the rock, with a doming high ceiling. Below, a tidal race runs from the Twll, a small gateway between the landward and seaward sides of the island. The water shines bright, turquoise, the light dancing patterns on to the rock of the roof. Every year, around eighty pairs of kittiwakes build their nests and raise their young on the edges of the cliffs and caves of the island before heading offshore again. One year, the birds built their nests, but they sat empty and no eggs were laid. Both birds would be on the wing or foraging all day; there was not the usual back and forth as one of the pair brought food to the other who sat brooding the clutch. In 2017, the UK's kittiwakes were declared vulnerable on the International Union for Conservation of Nature (IUCN) Red List of Threatened Species, with the population in steep decline. (St Kilda alone has lost approximately 90 per cent of its breeding pairs since 1970.) Although the decline of any species can be complex and multifaceted, with ecological models being constructed in order to try and explain the underlying causes and, if possible, reverse the plummeting numbers, in the case of the kittiwakes, there are two likely culprits.

The first is the fishing industry. Like cetaceans and seals, the birds can become tangled while feeding and drown in the nets.

But there is another danger. Historically, the kittiwake would have been in competition with other marine life for the sand eels that predominantly make up their diet. Now, they also have to compete with trawlers, which net and harvest great swathes of their food. The fished sand eels are not directly for human consumption. The majority are to feed salmon and hake in fish farms, or pigs.

The second factor in the kittiwake decline is anthropogenic climate change. Our seas are warming. As they warm, the ocean currents, which used to be seasonally predictable, are starting to shift. This leads to plankton blooming in different places and at different times. With this small shift, we see vast changes in the food chain. When planktonic blooms move, or are earlier or later than expected, the sand eels don't thrive, and so the kittiwakes go hungry. Perhaps the gulls that I encountered and observed on Ramsey Island had returned to their breeding site so undernourished that they simply could not produce their clutches of eggs that year.

I was nine years old when I saw my first whale. It was dead, washed up and misshapen, gravity holding it all wrong, out of its context on an exposed south-westerly beach. Underfoot, I felt the crunch and squelch of seaweed, rank and rotting, the top layer crisp from the sun, the underneath dank. Detritivores and sand fleas leapt from my path. The whale's skin was thick and dark, somewhere between slate and obsidian. Its head was a voluptuous curve, its tail torn by the rocks, with ragged flashes of red. Otherwise, it looked whole. It was fresh: necrosis had barely started to set in. It was a pilot whale, and I don't know why it died. Perhaps it was natural, part of the cycle of the ocean, perhaps not. But I felt a reverence in its presence, a pull that I could not fully explain. People wearing sterile gloves came to put plastic sheets around the whale and take it away for autopsy. As it was moved from the sea for the last time, I felt a resounding sense of sadness. I'm sure the detritivores did too: the largest of feasts had been set before them and now the table was being cleared before they could eat. The whale would be chopped up, its bones examined for potential aeration, blubber samples taken, internal organs scrutinised. I wanted to know how it had died, but, more than that, I was curious to know how it had lived. Where it had been, what it had seen. I wanted to leave the shore behind, to venture into a world of motion, dictated by the tide and wind, the world of the whale. To see what was over the horizon. I wanted to go where the whales go. This was the first time I remember feeling that tug away from home and out to sea. And, with each passing year, that feeling grew. It was a restlessness. As each seabird left its perch, as the seals started to slip away, as the swallows left our skies for a warmer continent, I felt as if they took a piece of me with them.

September meant school. It had started well. In my primary school, most of the day was taught in Welsh, the alphabet was *a*, *afal*, *b*, *beic*, *c*, *cathod*, *ch*, *chwilen*. I learned to play the harp there – a great full-sized pedalled instrument that sat on the edge of a small stage, cloaked in red. One wet and windy break time, I had sidled up to the instrument, pulled away the cover to reveal gleaming varnish, and started to pluck at the strings until my fingers blistered. The school arranged lessons. I was so drawn to the soft rising music of the country. Secondary school was different. Learning was more focused, and more sterile, and increasingly I found myself staring out of the window and gazing towards the sea. I remember Careers Week. Long afternoon talks from the marines were pitched to the children as adventure, the inevitable violence concealed. Vocation: plumber or electrician. Then, one by one, we were sent to the careers' officer in a small room, painted a sickly peach. I had never seen her before – she definitely didn't work in our small school year round. I sat, crossing and uncrossing my legs, the discomfort of my mind becoming physical.

'So, have you thought what you want to do when you finish school this summer?'

The truth of it was, I had thought about what I wanted. The problem came in articulating it. It was a *hiraeth* of sorts – something I was longing for, a longing that had been growing within me for years. I wanted to set off on an adventure – a naturalist with a notebook, taking measurements, documenting, drawing – and see what the world would show me, with space to think. Like Darwin on the *Beagle*: a voyage of scientific discovery full of iguanas and finches. I wanted to feel intrepid, to experience the rush of nerves as you push your own boundaries, like Amelia Earhart flying over the Atlantic, the first woman to do so. I wanted an odyssey of my own. I wanted to explore, to fill the edges of my mind with the edges of the world, to come back with a hundred answers and a thousand more questions. I wanted to ride the wind and swim in other seas. I wanted spice, colour, heat, cold, long days, short nights, aching body and racing mind. But, right now, I was just a seventeen-year-old from the coast with a handful of change from pulling pints in the pub in the evenings.

'So you don't know, then. How about . . .'

She flicked through a file, laminated cards of possible futures.

'Art therapist?'

I tried to explain the kind of life I wanted.

'Unrealistic.' 'But how would you do that?' 'That isn't a job.'

Outside, the warm wet storms of autumn were turning to the cold depths of winter, the windfall leaves blown away, branches bare. The puke-coloured room felt hot and close around me, my mind reeling with, 'You can't.' I stood up to leave.

Winter. Sharp winds battered the coast. The boulders framing the beaches would be lifted and hurled across the sand by the strength of the swell, the grains shifting to both cover and reveal. The auks were all at sea, rafted on the surface, letting the weather roll under them towards the coast. Herring and mackerel long gone. The common dolphin were nowhere to be seen, foraging far away in warmer waters, most likely Biscay, if not further south. A few seals still paddled about in the water, mostly youngsters who hadn't yet ventured further afield. Porpoise still surfaced with a puff, feeding, feeding, always feeding, if you knew where to look. I walked the coast path obsessively. I spent less and less time at school, slipping away whenever I could, over fields and down the old fisherman's track to the water. As I navigated the smooth dips worn into the cliff by years of footsteps, I pondered them curiously. Everything we do leaves a mark, it seems, in one way or another. I wondered what trace I was leaving. What mark the landscape had in turn left on me.

After nights at the pub, waiting tables, my journeys home became increasingly meandering. I'd often end up at the beach, a lone soul in the quiet dark, sitting in front of the empty lifeguard hut, gazing over a cold, inky ocean. The flash of South Bishop Lighthouse a metronome to my floating thoughts. The world seemed to be in greyscale. Pewter sea, ashy piles of spindrift on sombre beaches, the foam snatched by the wind and whirled upwards in vortices, the closest we really got to snow. Black tar lichen on the rocks marking the splash zone, thriving in the salt. Bracken, heather, bramble, their colours dulled and muted. It rained often in winter, great lashing sheets that would soak you in an instant, or a soft pervasive drizzle, warm and cloying. People stayed indoors, and those who lamented the crowds in summer were now hit with the loneliness of the off season. The sea was a comfort; oblivious to my frustrations, it welcomed me every time. The winter seas were rough to swim in. The swells swept me up, sending me tumbling towards the shore, snot streaming, hair in impossible knots, invigorated. These icy plunges were as settling to my spirit as the warm fire that followed was necessary for my body, when I would curl up with a book and lose myself in a world of adventure. In those last years of school, I was filled with an energy that lit me up and felt like it would come bursting from my skin, but with absolutely no direction. There was no north to my compass. No planned route to follow. I had been living a life of childlike simplicity, day after day exploring caves and coves, poring over rock pools, riding swells, dangling from clifftops. Now I had no idea how I was supposed to fit myself into adult life, and there was a growing pressure of expectation.

One morning, I left the house early, on the familiar path to the sea. Every step had an ache, not the stumbling feel of new muscles, tripping and striving, that I'd felt on my early solo journeys. This time it was a heaviness, a nostalgia. I passed the beach, the site of my first swims, but wandered on with little pause, taking the long arm out to St David's Head. My path was a peregrination, over the drystone dyke, past the hut circles ringed into the ground, from a time when people lived closer to the sea. Out on to the headland proper. There I could feel the breeze on my face, warm and wet, the tropical maritime blowing over the sea, all the way from the Atlantic islands. I sat with my back against a boulder, peppered with lichen that pricked slightly. I closed my eyes. I could feel it all around me, every inch of this landscape, this seascape. The turn of the tide, the echoes of a booming swell in caves and crevices far beneath me. The islands, the shape of the coast. My cottage. My harbour. I knew them all. I loved them all, so deeply it resonated in my chest, physical emotion. There was a part of me that belonged here, and that part would always belong here. But also a part of me that needed to wander. Stone against my back, sea before me. This was a place of edges. The western edge of Britain, the edge of my world.

The community, my home, is considered peripheral these days in the context of the country as a whole. It didn't used to be. Like the ocean, we humans are never entirely still, nor should we be. We hunted and gathered, and then we settled. We made the irreversible switch from existing in the environment, where our behaviour and our lives were controlled by the ebb and flow of the seasons, to trying to control that environment. Six thousand years ago, when people started to settle here, the water was everything. The sea, the rivers – they meant connection. As this community by the sea developed, there was trade, immigration, emigration and invasion. Wool, butter, cheese and cloth were exchanged for wine, iron, honey and salt – with Bristol and Devon, but also Ireland, France, the Basque Country, as far south as Portugal. Pembrokeshire's importance stretched all the way to the Mediterranean, and was recognised by the Romans. The very headland where I now sat was known to them as *Octopitarum promontorium*, warning of the dangers to ships posed by Ramsey, and the rocks and islets of the Bishops and Clerks. The headland was an important navigational aide, a prominent marker on entry or exit to the Irish Sea. Norse raiders left their mark in toponyms. Skomer: *Skalm*, side of a cleft. Ramsey: *Hfrans Island*. The South Bishop was known by a less clerical name, *Emsger*, isolated rock. Archaeologists have used stable isotope analysis to reveal a heavily transient population. The bones found in Pembrokeshire soil show origins from old Welsh, old Irish, Mediterranean, Norse, Flemish and Norman.

In the mid seventeenth century, there was movement across

the Atlantic, as the Quaker populations, under persecution in Wales, found home on the shores of the New World. In 1888, they returned from the island of Nantucket, illuminated with the profits from selling whale oil to burn in lamps. As they lit the country, they also built the port of Milford Haven, with hopes of establishing a whaling presence on the shores of Pembrokeshire. Ships were built, right here on this coast – sloops and smacks, schooners and brigs. I wish I could have seen it: the sawing of wood, land to sea, as trees became planks, planks became hulls, as canvas was hoisted to catch the wind and ships became their own entities on the water. Tenby Harbour was filled with three-masted fishing smacks, whose catch was an abundance of herring, until the stocks crashed in the mid 1800s. Oysters were dredged up the River Cleddau under sail, until they too went into serious decline in the final years of that century. I wonder if people then considered the sea a boundless source from which they could just take and take. And what they thought as the catch dwindled day by day, net by net. The sail trade in Pembrokeshire ended with the rise and rumble of trucks and lorries, around the 1930s. Fishing continued a little longer, by swapping sails for steam. There is still commercial fishing in Pembrokeshire, mostly pots – crab and lobster – but the scale is far smaller. The majority of the fishing boats are under ten metres now, and typically fish in inshore waters. Alongside fishing, there was oil. The port of Milford Haven is recognised as the largest natural deep-water harbour in Britain. The refineries that slash the skyline, built in 1957, with a sort of dystopian beauty, were constructed as oil tankers grew to 100,000 tons. Later, it would be 200,000. And then 300,000. Now, there is a proposal to build an experimental nuclear fusion reactor on the waterway by 2040.

The community on the River Cleddau and the people by

the sea were once considered the most connected, when connection was via water, not based on the strength of a Wi-Fi signal or proximity to London. Now, they are on the fringes, dormitory, next to, rather than being valued in their own right. The population here is growing older all the time. There is a net export of young people who go in search of opportunity beyond the tourism that now dictates the industry here, with its seasonal work and low pay, juxtaposed against house prices that are unreachably high. I have friends who have decided to stay here, and found ways to made their livelihoods. But I know many more who have left. Perhaps this is the same in many coastal communities.

One winter day while walking on the beach, I found a message in a bottle. It came from Guernsey, written by a group of three teenagers about to go on a night out at the end of summer, before they went away for university. They had drunk the bottle of red and, in the joyous spirit of three young people about to spend one of their last evenings with friends on a small island, had decided to write a message, telling the world who they were in that moment. They sealed the bottle with wax, and hurled it off the harbour wall. I received it months later. Who knew the journey it had been on before it came to me. There is connection via the Gulf Stream. The current of water that keeps serious frosts from our coast, keeps the climate mild as it kisses the peninsular, also touches shores in the Gulf of Mexico, a physical connection that stretches over thousands of miles. The water in which I swam had met whales – right, humpback – as they made their migrations north and south on the other side of the Atlantic. Increasingly, I felt it too, the call to migration.

A throaty cry interrupted my meditations on home and country, through past and into future, an anchor drawing me back to the present. My eyes flew open. There, right in front of me.

Black of wing
Night black. Feathers flash.
Stretching out, fingertips,
Peel the very air.
Red.
Bright red.
Beak and foot,
Flashes. Blazes.
Burns the soul.
Fire crow.
Soars.
High and higher,
Air thinner, dancing updraughts.
Wings tucked, tumbling,
Twirling, falling,
Fast and faster,
Until.
Quick.
A rush, a thrill,
And snap.
Wings unfurled,
To climb again.

Pyrrhocorax pyrrhocorax. Cornish chough. Red-billed chough. *Palores*, the digger. The sea-mew. Loquacious crow. Sea crow. The bird has many names, but for me it is fire crow. At first, there was just the single bird before me, the corvid that had captured my attention, but then the rest of the flock appeared. They danced on the wind, eight birds in total, likely four pairs.

They seemed to marvel in their own flight, in their acrobatics, just as much as I did, rising on thermals unseen to me, before tucking their wings in tight to their bodies, plummeting with the weight of gravity towards the clifftop. They would snap their wings out at the last minute, dare-devil, skin-of-teeth, before banking elegantly, and rising again. Their wings were ocean dark. Their beaks and their feet red like a lick of flame, a fire against the winter sky.

The fire crow has become a bird of legend. They say that when King Arthur died, he became a chough, the red colouring at beak and foot a sign of his blood, and that through the bird the king lives on, will never die as long as they fly. Fire crow flocked around Circe, the first witch, winging overhead as she walked the island of Aeaea with a lion at her feet. The bird was seen as a firebrand, capable of setting both crop and home alight with a touch. You won't find the fire crow inland, although perhaps in the past you could. No, this is a bird of the edge, of the fringe. They live entirely on the west coast of the UK, and the west coast of Ireland – the Celtic regions. Fire crow are plentiful in Pembrokeshire, inhabiting both the islands and the coastline. On Ramsey alone, there are usually around nine breeding pairs.

As one of their other names suggests, they were once common in Cornwall too. Over the last century they fell into decline. At one point, there was only a single successful breeding pair on the Lizard, that jut of land stretching out into the sea, the southerly tip of the country. After their last successful copulation in 1947, the chough vanished from Cornwall for over fifty years, only found on the Cornish coat of arms, standing proud, flanked by both fisherman and miner.

The problem for the chough was land use. We changed how we live on the coast. Fire crow like the weather-beaten, the saline. They dig for grubs in poor-quality grazed soil on the coastal fringes. When farming moved inland, away from the salt, when heath and scrub were abandoned for close-packed farming, where roaming was more easily controlled, the scrub grew too high and fire crow could not dig and therefore could not feed. As the birds became rarer, the value of their eggs increased incrementally with collectors, and so were stolen from the few nests that remained. In the predictable way of humans, the loss of the bird was mourned in Cornwall far more than its presence had been protected, treasured only after it was gone. And now steps have been taken to bring them back. In the past decade, the birds have successfully, naturally, returned to Cornwall, a pair nesting and breeding on the Lizard again in 2001. Now, the coastline is carefully managed to maintain their habitat there, and in Pembrokeshire too. Right here, on St David's Head, ponies graze to keep the scrub short. On Ramsey, sheep are shepherded across Ramsey Sound by boat to ensure fire crow can feed. Monogamous by nature, the chough build their nests on the cliffs, in caves, in the eaves of abandoned buildings on the coast. Nests are lined with wool, hair and thistledown. Once a year, eggs are laid, chicks fed on regurgitated grubs, before fledging and establishing their own nest sites elsewhere. They live their lives over the sea, next to the sea, but they are not of the sea. They live on a thin divide, over a shift of matter. To me, that day, they were a bright, burning beacon. A blazing sign across the winter sky. My steps home rang with a surety that had been scarce for months.

That night, I lay in my attic room. The lighthouse beacon swept through the darkness, stretching over the coast, and out across the water. That light. It was no longer my faithful nocturnal companion, but a lure, calling to me, calling to me with every sweeping rotation. The rain thrummed, heavy and hard, the beat of a drum upon the slates, drawing me up. I peeled my back away from the reassuring warmth of the chimney breast, moved aside the inevitable pile of books, and dislodged the cat that always seemed to find his way on to my bed. Bare feet over rough floorboards, light steps, trying in vain to prevent the creaking of the wood as I walked to my window. Water on the inside, as my breath pebbled into droplets, condensing on the glass as I pressed my forehead into the cool surface. Water outside, as the rain came down in a torrent. I could see it pooling as the soft earth became saturated and readying to flow down the lane and on to the road in a stream. I could just make out the great ash tree swaying in the winter wind, leafless branches whipped into a frenzy. There were no stars that night, the moon cloaked from view behind the dense cloud. All my life, I had lain sheltered as the weather passed, safe and warm, watching, listening, waiting. But the wait was over and the time had finally come. Now, I would stand in the night. I would venture out, and discover that hearing the weather pass over the rooftops is an entirely different thing from feeling the lash of the wind and weather on your body in the darkness. That being out there in the elements creates an entirely different space in your mind.

2

Sperm Whale

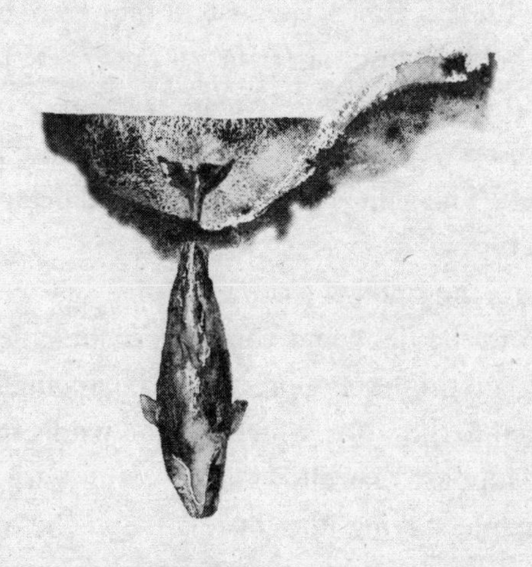

It was hard to tell the sea from the sky – the water was everywhere. Whipped into a froth as the waves hurtled towards us, a shower of freezing salty spray contrasting with the rain that fell so hard it burned. Only a slit of my face was visible under the hood of my oilskins, my eyes squinting, my hands bare, red and mottled as they clutched the wheel. The skin on my fingers wrinkled and puckered with all the water, tiny valleys in which the water could flow. I had no idea where my gloves were any more, not that they would have been much use in such a deluge. Wind, rain, wave after wave, indiscriminate roarers that crashed against the hull of the boat, showering past the spray hood that still stood valiantly. The whole crew were awake – of course they were, sleeping was impossible at times like this, hurled from one side of your bunk to the other, moments of free fall as the boat lurched and fell over the crests of waves. Wrapped up tight against the weather, we swapped 'would you rather' scenarios to keep the mood light, made up stupid jokes, the stupider the better, handing cups of tea up from the galley, which overflowed with rain almost the instant they were on deck. You had to keep hold of your mug tightly – never safe to put down, ready to hand quickly to a crew mate if the weather helm got heavier. Wet as it was, fierce and wild in the wind, I was glad to be above deck. The galley was the ultimate danger zone for me at times like this.

Down there, the smell of gas was all that was needed to send me flying on to the deck and vomiting spectacularly over the rail. Cooking anything was a precarious balancing act. You had to lash yourself to the stove, which swayed wildly on its gimbal, the pivoted support that allows it to swing with the motion of the boat while staying upright relative to the horizon. The idea is that while the stove swings, your food stays safely on the hob. To take something out of the oven, you had to take a chance, a minute or so where you would unclip yourself, open the oven, remove your tray of food and try to find somewhere safe to put it, while keeping upright, negotiating the dance of the wind and the waves as they threw the boat around. No surface that wasn't on a gimbal was truly safe, and only so much stability was gained through the friction of rubber mats. You needed to stay focused. One evening, I was making dinner while the wind blew thirty-five knots. A moment of indecision saw me hurled across the galley, almost in free fall until I came to a sudden stop, perched on the chart table, a tray of golden butternut squash and roasted garlic still miraculously held upright in my oven-mitted hands, rather than poured, burning, all over my face. The end goal was a risotto, and I have never been able to eat one since without remembering that moment and laughing. Opening lockers was a sport, as pots, pans and mugs would hurl themselves at you with uncontrolled vigour. Toilet visits were quick, with practised bracing, an arm or a leg against the wall to keep you in place. Bruises were inevitable, darkening the tops of my thighs the same mottled purple as the blackberries that used to stain my hands every autumn. In my roamings around the coast back home, I had moved through the landscape. Now, the seascape built, fell, hurled, roared and hurtled around me, dictating my movements with a Mephistophelian chaos.

A little black shape, a speckle of white at the base of the tail, flitted through the waves, a quick winged grace setting stark contrast to our laboured progress. Storm petrel, sea swallow, whipping over the water. I had always known the bird to be an omen of bad weather, storm summoner, the last bird home before the weather hit. Dance on, little storm bird. I pointed to it as it glanced over the sea.

'Hannah, I don't even know how you can still see,' a crew mate laughed, the rain lashing down that hard. We had only met a week ago, but bonds form quickly at sea. You are trusting each other with your life.

'Don't worry, I can't,' I shouted back with a laugh. A half-truth. Even through the driving, stinging rain, the chaos of the water, I could still make out the entrance of the first breakwater, the point where I would hand over to the skipper to take us to dock in the shelter of a harbour at Bonavista in Newfoundland.

We had only been at sea for a week, and we'd been planning for three, but we were running reefed before a storm, sails tucked down to the boom, the area reduced, a new tack and clew established to handle the strength of the wind. It had built to a severe gale as we approached the coast, but would become a violent storm or hurricane by the end of the day, with the force of the North Atlantic behind it. The dock was slick with water and hailstones as we arrived, storm-blown, within the hour, cold-stiffened hands working lines to safely tie up as the weather continued to blow. The harbour was full of brightly coloured fishing boats, only a handful of yachts braving this exposed Atlantic shore. For five hundred years, this was an incredibly busy fishing port on the North American coast. In 1497, a man known as John Cabot, an Italian sailor in the employ of Henry VII, sailed from Bristol in search of new land; at Bonavista, despite evidence of the presence of the Beothuk people, he staked a claim for Europe. The Beothuk were subsistence hunters, trapping ashore, and hunting seals, seabirds and fish on the coast, avoiding spawning periods in order to preserve the fish stocks while supporting themselves. However, the cod were so abundant that the seasonal European outposts became permanent settlements. For the Beothuk people, this change was devastating. When the posts were temporary, they had been able to coexist with the Europeans by avoiding contact, moving inland when they were at their fishing stations. When the colonists were there all year round, the numbers of Beothuk dwindled into extinction as a result of persecution, introduced disease and starvation, as European interference altered migration routes of caribou. The last Beothuk woman died in St John's on Newfoundland in 1829.

The cod continued, net after net, haul after haul pulled from the ocean. Innovations after the First World War led to longer expeditions to the nearby Grand Banks, a place of astonishing abundance. The new ability to process the catch at sea minimised time spent ashore and maximised fishing time. It is hard to be definitive about the impact these new practices had on the size of the cod population, as there was never any baseline data taken for the Atlantic cod. Estimates have been made, but all we have for sure is the catch data, which at its peak in 1968 saw 810,000 tons of cod landed. By 1992, the percentage of mature adults with the ability to spawn, known as spawning biomass, was estimated by fisheries scientists to be as low as 1 per cent of the peak population. The fishermen still went out with their nets, but they were now pulling up a smaller and smaller yield from the water. This dramatic crash led to a moratorium on cod fishing, and the biggest employment lay-off in Canadian history. In Newfoundland alone, around 30,000 people who had previously been entirely reliant on the sea for their livelihood found themselves without jobs.

The rain had turned to hail and the pontoons were slippery, requiring careful footing as we secured the lines and springs to the dock. Ahead of us, the harbour office stood tall, painted bright blue, well kept. All the buildings were the brave colours of a crayon box, straight-lined, tongue-and-grooved or shingled, giving no hint of the town's tragic history of colonialism, greed and mismanagement. Slightly land-sick, our heads still at sea, we all soon dispersed on separate wanderings around the island, all craving a moment or two of solitude after confined living. I found myself in a snug café as the rain hammered outside; once settled with warm coffee and sweet crêpes, I peeled off my wet layers. The owner asked where I was from, and there was a delightful moment of connection as he told me his wife was also from Pembrokeshire, that they had met on the cliffs there years ago, before she emigrated to Canada. Warmed with solidarity, coffee and a moment to catch up with my own thoughts outside of the practised routine of watch-keeping, I set off back to the boat. Earlier, I had walked inland across a decked trail by a lake. This time, I chose the beach.

The sands were littered with capelin, the silvery bodies shining against the sand. It was the first time I had seen the fish, and I was fascinated. At first, I thought there must have been some terrible stranding, a pollutant in the water that had caused the little fish to wash ashore. I didn't realise that this is actually a part of their ecology. To spawn, mature capelin hurl themselves out of the water and on to the shore, walls of fish washing up in the swell. The females lay their eggs in the sand, to be fertilised by the males, before the majority of the adults die, stranded and unable to make it back to the sea. The fertilised eggs lie buried in the sand and stones, as they were now, somewhere beneath my feet, waiting to hatch between a fortnight to a month later. Once hatched, the spawned capelin linger temporarily in the sand, before a mass exodus to the sea. In the water, they begin to feed on plankton, moving nutrients through the food web. They themselves form a crucial part of the diet of whales, seals, seabirds, squid and Atlantic cod. These tiny little silver fish are so small and delicate-looking, and yet are a part of the puzzle holding an ocean ecosystem together.

I looked beyond the beach, over the rocks to the harbour, at the boat I was sailing aboard, my home for the rest of the month. Navy-blue fibreglass, a squat double-ender with a rounded canoe-stern transom. A Valiant 40. It wasn't particularly exceptional in design, a production boat that you can find in marinas around the world, although I will always have a soft spot for them. No, it was not her lines that made this boat notable and gave her a special place on the ocean, but her purpose. *Balaena* was the sailing boat of Professor Hal Whitehead, a prominent and prolific biologist who had pioneered sailing cetacean studies since the 1970s. He and his students had sailed this boat more than halfway around the world, collecting data on whales and on the seas. On this particular voyage, Laura Feyrer, a PhD candidate at the Whitehead Lab (now Dr Feyrer), was collecting data for her thesis on northern bottlenose whales (*Hyperoodon ampullatus*). She would conduct four field seasons aboard *Balaena* in total, but this was her first. We were carrying out acoustic transect surveys, sailing the boat along a set course while trailing a hydrophone, an array of underwater microphones. We also had a dedicated observer on deck spotting whales and recording environmental data. Dr Feyrer's area of interest is the structure of populations and their geographic distribution. Her findings have been and will continue to be instrumental in assuring that any conservation strategies to preserve these whales are based on where the whales are actually found, how they use their habitat, and how threats affect them through the different stages of their lives. Alongside being a brilliant scientist, Dr Feyrer is also a mother, and has managed to balance her rigorous field schedule with parenting her young daughter. Sailing with her, I felt I had made it. I had found my north, the area of life into which I wanted to pour my passion.

The journey to get here had been a tough one at times. When I left school at eighteen, I got a job on a boat offering wildlife tours around the Pembrokeshire islands. I learned a lot very quickly, and I also learned a lot about what I didn't want to do. I haven't found myself in many environments that aren't male dominated, but this was one of the most extreme. As a young woman, you work twice as hard for half the opportunity. You prove everything, and prove it ten times again. Physically you have to do everything your male counterparts can do, even though your body is biologically different. There is no room to make mistakes. You can't complain. Any break, any opportunity you get, you have to be grateful for, knowing it could be taken away at any time. You don't talk about it. You just get better. And better. And better. Until you are exhausted. Working outside in the salt and the sun all day, every day, with scarce or non-existent breaks, I contracted a kidney infection from dehydration; despite being diagnosed by a doctor, I was told to turn up at work or I would be fired. At the height of summer, I worked seventy-hour weeks – boom in the summer, bust in the winter when the tourists vanished. I was considered free-lance – no sick pay, no security – and yet when I tried to crew for other companies as well, I was told this was unacceptable. I remember screaming at the sky on my bike rides home at the sheer unfairness of it all.

Historically, women in Britain were considered bad luck on boats, the idea being that their presence would whip the seas into a fervour. This was partly a manifestation of the suspicious nature of sailors. No whistling at sea. No voyages beginning on a Friday. Women bring bad luck. But it was also misogyny. While every element of our society is governed by patriarchal bias, at sea it is so strong that women have been excluded almost entirely. There are accounts of the wives of both merchant and whaling captains who went to sea with their husbands to avoid years of separation and found themselves confined to the captain's quarters, with an area of the aft deck that they could use to take air. Women making the crossing on ships between Britain and India in the days of the Empire were given very strict instructions on how to dress and conduct themselves on board, lest they elicited a wave of temptation, disrupting the orderly running of a vessel. They were always kept separate from the crew, in what must have been the most isolating and bizarre experience. There were a few women who went against the grain, boarding ships not as wives but as sailors, dressed as men to join crews. Many had very successful careers, although they had to leave if their gender was discovered. In all of these peculiar contradictions, maybe the strangest is that we gender our vessels: ships are *she*. The pronoun comes from the curves of a hull, and how a boat carries her crew, womblike, only in this case the water is kept out rather than within, and delivers them safely to their destination. Wives, sailors, stowaways – the reality is storms will come and storms will go, no matter who is at the helm. You just need the strength and skill to sail through.

This attitude towards women is dying, but it dies hard. Every female sailor I speak to voices the same or similar experiences of prejudice or outright misogyny, and we've all had to find our own way of navigating the culture. I don't really know what the best route to take is, and I often find it hard to address inappropriate behaviour in the moment. I have worked in a boatyard that had no women's toilet, and was told I could not use the men's. There are the predictable comments. I get patronised when ordering or collecting parts. I still get stared at when I am bringing my boat in alongside a pontoon, and in every single boatyard I find myself in. I leave conversations going over what I wish I had said, torn between self-preservation and furthering equality.

I am searching for a balance, to not work myself to exhaustion, these old harsh lessons ingrained. And yet, even during that first job as a tour guide, I loved my days on the water, watching the tides turn as the hours did, being able to share my love of marine life, and learning how to handle a boat through some of the trickiest tidal waters in Britain. At the end of one particularly long day, a father of daughters slipped two bits of paper into my hand as he got off the boat. One was a message saying how important it was to see a woman working in this environment and how much he appreciated his daughters seeing me there. The other said that too, in the form of a fifty-pound note. There were other people who buoyed me up during those early years too. One was Ffion Rees, the person who had hired me as a tour guide and shared her wisdom and experience. Having a paid job allowed me to save a little money and start paying for my qualifications. Ffion gave me my springboard, and has since gone on to run her own operation, largely staffed by women, with an emphasis on conservation and collaboration.

I also started volunteering at a sailing school, learning the ropes in something a little larger than a dinghy. With the help of the family who ran the school, I applied for funding so I could add to my qualifications, starting to put some pieces of paper to my name. Sailing qualifications are expensive, and the cost adds up very quickly. Getting started is often the hardest part, but I was determined, I asked questions, met people, made connections, found a way. I was also privileged – as a well-educated cis white woman with an extremely encouraging family. There were a lot of unpaid opportunities. I made sailing work by waiting tables, pulling pints, shifts in a shop, whenever I was home. I painted and sold artwork. I know how much it costs to keep boats running, and I was glad to be gaining experience and learning before I was qualified enough to get paid. I could just about support myself, but the system of unpaid work excludes many and it puts a limit on diversity.

After my first year as a tour guide, I enrolled on science courses with the Open University so I could start to learn more about the sea. In my second year of work, I also volunteered with the sailing school. I would drive overnight between the two, sleeping in my car at service stations. The following year, I was invited to sail across a stormy North Sea, from the UK to Norway and back, on a ninety-five-year-old wooden boat, one hundred tons of wood sailing at thirteen knots on a stiff breeze. My partner, who I had met when we were both guides in Pembrokeshire, had gone on to work on this classic boat, delivering charter trips to guests, and taking disadvantaged young people out to sea. At the last minute, they were short of crew, so he asked if I wanted to come, in exchange for taking on the bulk of the cooking. It was the first time I had sailed away from shore. Although the North Sea isn't particularly wide, we didn't make port for around five days, and I had my first opportunity to spend nights at sea, taking watches, to travel by sail.

That first night I was handed the helm, I was so scared my hands shook. The North Sea has a particular roll to it, and it felt so entirely different from the Celtic shores where I had grown. As nervous as I was to have the tiller of such a large boat, I pushed myself to focus, to work through the uncomfortable, with the hope that one day it would become comfortable. There had been a wind farm to port, one of the many you find in this stretch of water, blades turning, and at this point I wasn't used to judging distances at sea, particularly at night. Although we were miles from the turbines, they seemed to me too close, pressing in the darkness. What's more, ahead was an oil rig, one of the 184 that pepper the North Sea, and it was flaring, a pillar of orange flame reaching into the night; all the while, huge tankers transited past at their accelerated speeds, while I maintained our course. The night seemed strangely dystopian, the rig like some kind of H. G. Wellian creature, striding from the sea on its stilted legs, ablaze, the tankers moving fuel and food around the world, the machines that underpin the reality of our modern lives, which are mostly hidden from view, out at sea. When I finally went back to my bunk, I found that waves washing over the deck had been leaking through the planking. All I wanted was to close my eyes and rest, but my bed was a puddle of seawater.

Although the wind built throughout the voyage, I grew steadier as we sailed north-east for Norway. On the morning we approached our harbour, I had my first experience of picking up land smells after so much time in the salt. The wind blew northerly, straight through the pines, the most beautifully fresh, verdant aroma meeting us as we came ashore. Throughout the whole crossing, I had only seen one minke whale, and a smattering of porpoise, although now I suspect that there had been more wildlife around, only that I had been so consumed with finding my rhythm on the boat, working in the galley and trying to make myself useful on deck to really focus on looking for whales. I stayed on board just shy of a month. Once in Norway, we sailed around the most stunningly beautiful archipelagos, racing other classic boats. The water was clear, and incredibly icy, a true freshening. From Norway, we sailed to Germany, and finally to Holland, before once again crossing the North Sea back to the UK. And then came the opportunity I'd been longing for.

The research trip on *Balaena* was the realisation of the dream I'd had since my childhood encounter with the pilot whale that had washed up on the beach in Pembrokeshire. I wanted to know, wanted to see, every single species of whale. I wanted to know where and how they lived, which habitats were suitable. I wanted to know how many there were. I now knew that the answer to these questions could be discovered by 'abundance and distribution' surveys. They sound quite basic, but they provide fundamental knowledge. The principle at the heart of ecology is to determine what lives where, and why it lives there. This is the information you need to determine whether a species is endangered or not, or to be able to implement any kind of effective conservation strategy. As much as I wanted to sail, wanted travel and adventure, I also wanted to contribute, for my journeys to play a part in protecting the seas. So, at the age of twenty, I emptied my bank account on a plane ticket to Canada to sail on *Balaena*. And now here I was on the other side of the Atlantic, on a scientific research expedition.

I spent my days climbing masts to look for whales. It was like my early exploration of the treetops, but better, green leaf turned to blue sea as an ocean opened out underneath me. Even the smallest movements of the sea were amplified with height, but I dared myself upwards, ignoring the small shake in my hands, the small tell of protest in my body, and swung into the crow's nest. I never wore a harness, never attached myself to a halyard, even though a fall to the deck would have been disastrous. I was invincible with the armour of youth, buoyed by the confidence of how my life was unfolding. Sitting in my perch up the mast, my feet dangling out into open air as the boat moved on the wind, I felt as if I could absorb the motion of the world around me, that I was a part of all of it. I think the best sound I ever heard was the tall loud blow of a fin whale as it surfaced. A loud puff on a quiet ocean as the long pillar of whale breath reached towards the sky. I had found the offshore world of the whale.

I felt rather than saw my first sperm whale. It was in the pitch dark of night. A thick, isolating fog was twisting around the boat, around me, tendrils creeping down my collar, down the companionway and into the saloon. It blocked out the stars, the moon; I could barely see my own hand in front of my face. It was cold. Bone-achingly, strength-sappingly cold, stealing spirit as well as warmth. I was still fuggy from my short stint of sleep below deck. I had taken to sleeping in all my thermals, all my clothes, all my waterproofs and a hat, wrapped in my down sleeping bag in my pipe cot. All the layers I had, and they weren't quite enough. On good days, my boots would come off. On bad days they wouldn't. The heater had broken days before. The salty air had given everything a subtle damp feeling, and I longed for real, true warmth. There was a fruit net hanging close to my pillow. Before we left the shore, we had stocked it with apples, firm, rosy, crisp and sweet. I had watched as the apples bruised, wrinkled and started to soften in the net, diamonds crisscrossing into the fruit as it started to ooze out of its confinement. I'm not sure I was looking much better at this point, hands wrinkled from the damp fog, my hair in a lank, tangled braid. Freezing cold and nauseous, bags under my eyes, from lack of sleep and lack of food. It's hard to know when to eat on such a disjointed schedule, waking up every two, four or six hours depending on the boat.

Now, at the start of my next watch on deck, I huddled shivering under the spray hood. My nausea was rising. On my way up to my watch, I had grabbed one of the spongey apples from the net and was forcing it down, knowing full well I would likely be throwing it up soon. Better to throw up something than nothing. In that moment, I was feeling miserable, sick, cold, uncertain, and far from home and the warmth of a hearth at my back. I was sailing, hundreds of miles offshore, on the east coast of Canada in the freezing fog. Why, for all my love of the ocean, for all my love of exploring and the unknown, was I putting myself through the discomfort? We weren't even sailing, but drifting, hove to for the night, on the lookout for icebergs. The boat was almost at rest as the backed headsail tried to pull us one way, the main the other, *Balaena* as close to still on the sea as she would ever be. There was nothing to do but stay awake and watch, both in the night and on a radar screen, in case icy shapes loomed their way out of the fog. I stood, apple bits rising, and leaned over the guard rail to vomit into the water. But, something had changed. I froze, as I felt instantly and sincerely that I was no longer alone on the sea. There was a dark shape next to the boat, as if land were rising from the water. But rather than inhaling the scent of earth, I felt a deep marine presence in the night. I peered further. Something lumpy, a soft glisten of water on an almost gelatinous form. Almost as if it took offence at my curiosity, the animal sent a wet cloud of breath spraying into my face.

> *Physeter macrocephalus*,
> Sperm Whale,
> Ridged from fin to fluke,
> A low humped dorsal,
> An island in the ocean.

A bird's eye view gives no indication of the scale of the creature, the proportions the Latin name alludes to. *Macrocephalus*, big-headed. Their skin peaks and troughs into valleys, a landscape to be explored. At the largest rise is the crater of the blowhole, placed slightly to the left on the animal's head. The bulk of the animal, the powerful fluke, the jaw, the teeth, are all hidden beneath the surface. I don't know how long it had been there. It was more likely that it had been logging on the surface, sleeping or travelling, rather than surfacing from a dive. I don't know how long it stayed after our encounter, whether or not it shallow dived immediately, as surprised by me as I was by it, but I did not see it again. I'm not sure I would even have been able to make out a fluke in the black of night. I started to laugh, feeling instantly warmed from within as I wiped the spray from my face. This. This was why I wasn't a thousand miles away, wrapped in a duvet, with only dreams of whales. Laughing at the freezing skies, wiping away the dripping cloud of whale breath before it froze to my face. This was one of the reasons why it was beautiful, for all the reasons that it was painful. Worth every apple vomit, worth the isolating fog, worth the bruises, the burns. Ever since I became fascinated with whales as a child, the sperm whale had occupied a central place in my imagination. But there was a profound difference between loving an idea based on watching documentaries and studying pictures in field guides, and meeting that idea as a creature in the flesh, on a cold, foggy night, just the two of us.

Sperm whales are the largest of the odontocetes, the toothed whales. They are the species of whale in every cartoon drawing, an emblem of the ocean giants. You cannot understand the shape of the sperm whale without looking at them underwater. Their bodies, the lumpy and awkward hillocks you see on the surface, become graceful curves as they glide through the ocean. They must be understood in their context. Sperm whales are mammals like us. We both evolved from fish, lobe fins gradually becoming legs as tetrapods moved away from the water. The marine mammals then evolved to go back to the sea, but we still share a major physiological feature: we both need air to breathe. The deepest a human has dived on one breath is 214 metres. Sperm whales can regularly dive to 2,000 metres. For them, it isn't a great feat of endurance, of mastering the mind and the body as narcosis grips perception and the nervous system, but part of their daily forage. We both share aspects of what is known as the 'mammalian dive response'. As we hold our breath in water and begin to dive, we experience bradycardia, a slowed heartbeat. The blood vessels in our extremities constrict as our blood is channelled to our vital organs. The same is true for sperm whales, only they are able to withstand the effects for longer.

As an adolescent, when I felt frustrated by life, I would go to the water, one particular spot, a sheltered cove where the walls of rock formed an underwater grotto. It's not a beach, more a large stack of stones at the bottom of a cliff, and, when I arrived, I would carefully select a big rock from the edge. Something round, something big, but something that I could hold on to. I would walk into the sea, cold biting at my ankles at first. Deeper, feeling my own skin slip through that of the water. I would breathe quick and shallow on the surface, before ducking, diving down, descending. At one side of the grotto, a kelp forest swayed, brown and golden, tempting me into its complex midst. I would sit there, on the sea floor. It wasn't deep – two metres at best. I would look up, to see the light rippling from the surface, streaming down in perfect dappled rays. Everything would move around me, gently wafting on the tide. I would watch mesmerised as speckled plankton drifted in and out of the rays of light, caught up in their own dance. I would clutch my rock tightly to my chest and try and soak it all in.

Down there, I felt I could sense the vast expanse of the planet's water. The 'seven seas' is an outdated phrase. These days, it is common to split the water into five oceans: the Arctic; the North and South Atlantic; the North and South Pacific; the Indian Ocean; and the Southern Ocean. There are numerous seas, and how and whether you divide them is dependent on culture and language. Different patches feel different, have higher or lower salinity. They have different weather. Hot quick squalls in the Caribbean. The wild endurance of an Atlantic storm. The desperate chop of the North Sea that never sits right with me. The brackish quiet of the Baltic. The different shades of blue, azure to lapis, jewelled to dark. And yet all of these areas of water are connected. Sitting there in my small cove on the Pembrokeshire coast, I wanted to see them all, feel them all. I would try and reach with my mind to the very edges. But even while my mind wandered, my body began to protest. My heart beat slow and loud now. A plea. Air. Air. I would start to feel my pulse in my head. My ears rang as the pressure to breathe mounted. My lungs beginning to shout. To scream. Filling. Bursting. I would sit as long as I could, until every instinct forced me to release the rock, to glide upwards. My head would break the surface and I would drink the air down, sweet like honey wine, as the water streamed from me.

I wonder if this is how it feels for the sperm whale too, that first breath on surfacing. They don't need to pick up rocks, of course; their bodies are perfectly adapted for the depths. The sperm whale will, in fact, exhale as it dives. Their lungs empty, they are designed to be collapsible, an adaptation to the immense pressure experienced when 2,000 metres of water stand between you and the surface. I can't even begin to imagine it. The weight of the ocean on their globed heads, Atlas carrying the world of water. Their oxygen is stored in their skeletal muscles and their primary organs, with myoglobin as the main carrier for their oxygen rather than the haemoglobin in our blood, enabling them to have enough oxygen to function despite the physical demands of their dives.

Sperm whales can see underwater, but sound is the primary medium through which they experience their environment. As a sperm whale descends from the surface, the first 200 metres of their dive will be through what we call the euphotic, or sunlight, zone. Here, the light streams through the water, photosynthesis is possible, and sea creatures can rely on sight. As the whale descends further, it enters the dysphotic, or twilight, zone. Here the light is sparse and the water begins to grow darker. As the whale descends past the 1,000-metre mark, it enters the aphotic zone, the midnight zone. Here, the world of the whale is bathed in darkness, wrapped in an inky black as it hunts for cephalopods, the squid that predominantly form the diet of the adult sperm whale. Typically, sperm whales will begin to vocalise when the light starts to fade as they reach the bottom of the sunlight zone. When listening through a hydrophone, you hear these vocalisations as a series of clicks: regular, metronomic. The sperm whale is the loudest single animal in the ocean, or it used to be before human interference. Click. Click. Click. I have fallen asleep hearing this regular soothing lullaby through the hull of the boat. This sound. This was the sound that haunted Melville's hero in *Moby Dick*. Click. Click. Loud, sure, resonant. We can hear the clicks, record them through hydrophones, but we do not yet know the extent of the information that they transmit. It is commonly accepted that sperm whales will use these regular clicks to determine how far away the sea floor is. They dive at such speed that they would risk collision if they did not know how far away the seabed was. We think that through echolocation they create a three-dimensional sound map of the world around them, affording proprioception, and that the clicks are used as a method of detecting prey. We also know that these clicks are used to communicate with other sperm whales, and that the nature of the clicks varies from clan to clan. But we don't yet know what else they are saying.

There is an ambitious project underway which could prove pivotal in our understanding of their vocalisation and in the way we interact with the world of the whale. Project Ceti combines artificial intelligence, engineering, fieldwork and linguistic analysis in order not only to interpret the whales, but, for the first time in history, to be able to speak back. Whether or not sperm whales actually want to talk to us is an entirely different matter.

Historically, we have made life very difficult for the sperm whale. It is not clear when or where whaling began, as it is an ancient form of subsistence hunting that sprang up simultaneously in various different coastal populations around the globe.

Signs of commercial whaling can be traced back to the Basque Country, where, as early as the seventh and eighth centuries, towers were erected in the hills to look out over Biscay to sight *Eubalaena glacialis*, the northern right whale. Whaling spread and boomed with the industrial revolution. Whales were viewed as a resource, something that was ours to take from the sea to be exploited for profit. Whaling became an enormous factor in global wealth. In 1847, the whaling town of New Bedford, Massachusetts, was considered the richest town in America. Every species of whale was hunted, but right whales were especially prized, both for being easy to track, as they have a tendency to surface in close proximity to the site of their last dive, and for being easy to collect, as they would often float on the surface after being killed rather than sinking. Whale oil was a valuable commodity, rendered from blubber and used for lubricating machinery and in paints, varnishes, soaps and leathers. Instead of the majestic ocean creatures winging their way gracefully through the water, whalebone was extracted from their carcasses to stiffen hooped skirts and corsets. Sperm whales were also attractive to the hunters. Spermaceti, the waxy substance that is found in the cavities in the head of the sperm whale, created the brightest, cleanest-burning candles. Ambergris, a bile secretion from sperm-whale intestines, was much sought after by parfumiers, and to this day retains a high value, up to $40,000 a kilogram. It is either found in the viscera of the sperm whale, or occasionally washed ashore of its own account.

As I child I hunted the Pembrokeshire shores vehemently, looking for this treasure from the sperm whale. I remember searching the steep pebble-banked storm beach of Aber Mawr, on the north coast, on the off-chance some ambergris had washed ashore. To get there, I would walk through the woods, which were carpeted with bluebells and fragrant with the tang of wild garlic. The woods would fall away to reveal the bay, my eyes adjusting from the canopy dappling green light to the blue of the sea and wide sky. I would slide down the pebbles, the stones clattering and cascading with me. The descent was always easy; climbing back up after the search was far more difficult, as the stack of stones would sap the strength from my legs as easily as they did with the breakers, whose climbing waves would be sucked down between stones with a *shush*.

The sand here was dark, and in places it slid away to reveal the skeletons of the drowned forest, the soft dark stumps of petrified trees, land back to water. I would pick my way across the sands, eyes peeled and searching – early practice, perhaps, for when my eyes would scan the water, searching for whales during surveys. And there! My heart raced as I hurried over to the hard, waxy lump. Amber and ochre in colour, the large piece had obviously been on quite the ocean voyage, rough as it was at the edges. Delighted to find my treasure at last, my piece of whale vomit, I scooped it up and raced to show it to my mother. We scoured in books and online for the ways to tell if it was really ambergris. We heated a needle with a lighter, heard the soft bubbling hiss and watched as a white plume of fragrant smoke curled up. All the signs were there. I carefully shaved off a small piece, packaged it up with my intrigue and excitement and mailed it to be tested by a scientist at Bangor University. I went about the following weeks with my whale treasure at the forefront of my mind. The precious lump sat in pride of place on my windowsill, a piece of the wild world that I so wished to see, to explore, to be a part of. When I received a letter back I tore it open. 'We are afraid the specimen you sent us is not ambergris, but some unknown waxy substance.' To this day, I still have my waxy substance, buried deep within the book labyrinth that forms the architecture of my mother's cottage, a reminder of my childhood treasure hunt. I still hope one day to place a piece of the real thing beside it.

Commercial whaling continued, and whaling ships and methods advanced. The whalers' thoughts were on profit rather than exploitation. More. Faster. Better. Whaling was a gold rush, with no limits or regulation. In Western culture, we have viewed the seas as an endless resource. It is only in the past one hundred years that we have begun to see the effects of centuries of depletion and are taking steps to curb it. The first attempt at regulation came in 1924, though due to the difficulties of international collaboration stemming from cultural, political and economic differences, no accord was reached. In 1946, the International Whaling Commission, or IWC, was founded, with the purpose of regulating whaling. Friends of the Earth, a global environmental campaign, is highly critical of the IWC, stating that they acted in the interests of whalers rather than whales.

Initially, the IWC reviewed catch data from whalers every year, and then determined whether regulation would be necessary before the next review. An antarctic limit on whaling was set for the first time, along with an antarctic whaling season. The limit was 16,000 Blue Whale Units a year. A Blue Whale Unit, or BWU, consisted of one blue whale, or two fin whales, or two and a half humpbacks, or six sei whales. The quantification of this unit was based on how much oil could be rendered from these creatures. Initially no grey whales, humpbacks, or right whales were to be hunted. These measures were to safeguard the future of the resource. The IWC lacked any enforcement, and instead passed responsibility to each individual nation to adhere to the regulations. In 1949, the fishery opened for humpbacks in Antarctica, with a quota of 1,250 whales. That year, 2,117 were landed.

By 1965, the largest creature in the ocean, the blue whale, had almost disappeared entirely from Antarctica. Change came in 1970, and a chance for the future of our planet's whales. In America, whaling was falling out of favour in public opinion, and an environmental movement was growing. All commercial whales were added to the US endangered species list, and in 1972 the Marine Mammal Protection Act was passed. This act made any 'take' of a marine mammal – defined as any hunting, kill, capture or harassment of a wild marine mammal – illegal, and their products could not be imported or exported. This proved to be the turning point not only for America, but also for the IWC. The call to protect whales was still gaining momentum and a moratorium was proposed. Although it was rejected due to a failure to reach a majority vote in the IWC, quotas for both sperm whales and minke whales were reduced. The US continued to sanction whaling in their own country, while banning fisheries importing from countries whose operations were seen to undermine global conservation principles.

The role of environmental groups became particularly significant when a total of five million members, spanning twenty-one organisations, began a boycott of all products from Japan and the Soviet Union due to their whaling practices. The impact of all of these individual actions was incredibly powerful. Although a 1974 call for a moratorium was deflected, quotas were set that began to consider the ecological impact of the removal of individual creatures, based on their species, size and gender. Finally, thirty-six years after its conception, in 1982, the IWC called for global whaling to stop, with the exception of aboriginal hunting. The initial moratorium was set for three years, but thanks to repeated extensions, it is still in place. Japan continued under the banner of research, sanctioned by the IWC until 2019; they then withdrew from the committee and got to rename their research boats as whaling vessels. Russia has maintained a practice of deliberately landing illegal catch, falsifying records and misreporting their continued whaling activities, and yet it remains a part of the IWC. Whales are still hunted elsewhere too, in the Faroe Islands, Norway, Iceland, Greenland and South Korea, among other countries. Commercial whaling may have been greatly reduced, but its damage has been irreparable. Several species were nearly hunted out of existence.

In my days sailing Biscay, I have heard and seen the loud towers of breath blown skyward by fin whales, *Balaenoptera physalus*. I have watched the gentle surfacing of *Eubalaena australis*, the southern right whale, from Mouille Point in Cape Town. But I have never seen, nor am I likely to see, *Eubalaena glacialis*, the northern right whale. The species was plentiful at the point when those early Basque whalers in Biscay began hunting, but in 2018 the species was declared critically endangered by the International Union for Conservation of Nature, the IUCN. In that year, they listed fewer than 250 mature adults left in existence, all clustered in an extremely isolated population. While I was sailing the cold Canadian waters on *Balaena*, I was constantly hoping we might encounter them.

While the ocean giants struggle to recover, we are encroaching further on their environment in new ways, including through environmental pollution and underwater sound pollution. Of all global cetacean species evaluated by the IUCN, twenty-two species, containing thirty-eight subspecies or populations, are considered critically endangered, endangered or vulnerable. This number is likely to be higher, as nine species with two subspecies are listed as data deficient, due to lack of study. And, while we no longer reduce the worth of a whale to a Blue Whale Unit, we still run the risk of valuing them for what they do, rather than what they are. Even within the environmental movement, the conversation about whales is still often couched in financial terms. The economic value that whales add to society has been calculated. Now, instead of the profits from making hoop skirts, corsets and brightly burning candles, we are looking to value whales as repositories of carbon. A great whale can accumulate thirty-three tonnes of carbon dioxide in its body over its lifespan. When the whale dies, it is likely to sink into the deep ocean, where that carbon will be held in its bones. Live whales help to sequester carbon too, through their poo. Their faeces act as a fertiliser in the surface waters, helping the bloom of phytoplankton, which produce over half of the world's oxygen, exchanging it for carbon dioxide. It is estimated that the number of sperm whales alone killed as a result of commercial whaling has caused two million tonnes of carbon to remain in the atmosphere. But, for me, their true value is beyond economics, beyond science, even sometimes beyond words. Who would want to live in a world devoid of these majestic ocean giants?

The first time I saw a harbour porpoise – the first cetacean I had seen, before I even knew what a cetacean was – my mind fell quiet, thoughts held aside in a soft meditation. The world did not stop. There were still dishes to do, laundry to fold, places I had to be. But in that moment, I could not leave, and sat, transfixed, watching these creatures move through the world. The sperm whale in the freezing night, at the side of the boat, affirming my place on the sea. At a base level, a level that is somehow both more simple and more complex, I know that time spent in their presence, breathing the same air for a fraction of time, is a privilege. It inspires both awe and wonder, respect and reverence. That moment of connection, when we both share the same space – two creatures so vastly different from each other in many ways, less so in others – until the whale dives, with that lilting fluke raised towards the sky. Sometimes, something is just beautiful. You don't know why, it just is, and the world is better for the fact that it exists.

Sperm-whale societies are matrilineal. This means that their societies are founded on kinship with the mother, the female line. The females stay close to where they were born, while the males range in the ocean throughout their lives, typically heading to colder water, to find bigger prey, so they can grow larger and compete for mates. The females come into sexual maturity around the age of nine or ten. They have a gestation period – when they are pregnant – of fourteen to sixteen months. Once they give birth, the calf is suckled and fed predominantly on milk for an average of two and a half years, but females are recorded suckling young up to the age of thirteen. It is thought that this prolonged nursing period is for the benefit of the mother as well as the calf, as it gives her time to recover from pregnancy and birth before she becomes reproductively viable once again. The fertility of the female is greatly reduced around the age of forty, and yet she continues to be an important member of the group after this, and will live into her seventies. She does not become invisible to society, instead she becomes part of the circle of females collaborating to raise the community's young.

Sperm whales are slow to reproduce. Each individual whale will have only a few offspring and will invest greatly in parental care. This 'life-history strategy' means the sperm whale is described as a 'K-selected' species, as opposed to an 'R-selected' species, like rodents, which have a large number of offspring but do not care for them for an extended period of time. Raising a sperm whale is a community effort, as the mother cannot care for her calf alone while also feeding herself. Although the sperm-whale calves are born with a high level of development, appropriate for the complex environment that they find themselves in, they do not have the deep diving ability of adults. Studies have shown that while sperm-whale calves have the physiology for a deep dive from the age of one, they continue to nurse. This suggests that perhaps the prolonged nursing period allows the calf to develop other skills necessary for a successful dive, such as being able to recognise and successfully catch prey, and potentially to communicate with other sperm whales during the hunt. Left alone on the surface, while the mother dives for food, the calves are vulnerable to predation, by orca, sharks, false killer whales and pilot whales, so the other females in the community take on the parental role of protection while the mother is in the deep. Suckling mothers will feed calves that are not their own. If the group is threatened, or a threat is perceived, they form a shape called the marguerite formation. The calves are sheltered at the centre, and the adult whales crowd around, heads facing in, their powerful flukes facing out to beat and swipe at any approaching predators. They know they are safer as individuals if they work together as a group.

One summer, aboard a sailing boat in the Azores, I sat in the crow's nest perch, looking out over the volcanic islands, skin warmed by the sun. I watched as newly surfaced mothers replenished their bodies with air, calves charging up to them with seeming delight to be reunited. The waters around the Azores are thick with youngsters in August. Observing these whales, it's hard not to project human feelings on to them. The joy of being reunited in person with loved ones after time away. At that point, I hadn't seen my own mother for months, nor would I for months to come, as she largely remains on the Pembrokeshire shores. It is hard to spend time with the sperm whales without pondering my own matrilineal heritage, how things change through time, how information is passed along as an inheritance.

My maternal grandmother recently gave me a locket to wear at my neck – round, gold, scratched and dented. The front is engraved with swirling symbols, the back looks like the metal has been hammered, but I know the truth of it. The locket was a gift from my grandfather to my grandmother. It has pictures of the two of them inside, smiling in front of the old apple tree in their garden. The dent happened when my grandmother accidentally sat on the locket, an imprint of her pressed into the metal. She gave it to me shortly after my grandfather died. I was sailing half a world away and didn't get to see him as he faded from life. I couldn't be with my family, and, like many others in the time of the pandemic, they could not physically support each other at the small funeral. Now I hold it close, the two of them, as if they are together, for the last time. My grandmother told me I should get the dents hammered out, but I won't. They feel like part of their story, a moment captured in the metal.

My maternal grandmother was born in the Black Country, to a working-class family, in a terraced street of coal dust and the tang of the foundry. They say the days were black with dust, the nights red from industry. She talks of pigs at the bottom of the garden, growing vegetables anywhere and everywhere, every space used, so that they would have enough to eat. She left school early, she had to work, her family needed the money. She says she always wanted to be a teacher. Instead she became a secretary, out of necessity. She married a policeman in her early twenties, the age at which I was searching the North Atlantic for northern bottlenose whales and being sprayed with the breath of sperm whales in the night. Sixty-five years later, they still walked down the street holding hands, never spent a night apart until he died. She never learned to drive. She does not own a passport and has never travelled outside of the UK. Her world has been a microcosm stretching to the coastline of the British Isles and no further.

For a time I found it hard to talk to her. She stopped writing back to my letters, embarrassed by her spelling – unnecessarily so, as mine has always been appalling. I was often abroad, so the chances for phone calls were few and far between. I felt that our lives and experiences were too far apart. I was embarked on adventures she could never have dreamed of. I was studying cetaceans and chasing horizons. My nights were lit by stars picked out against the blackest of skies, my days spent navigating paths through Scandinavian archipelagos or weaving between fronds of coral on the other side of the Atlantic, sailing through turquoise seas. I was fully immersed in the present on my voyages and found it hard to explain my motivations or my plans for the future, as I rarely had any beyond the next moment.

It wasn't until a visit to her home on a hot summer's day that this changed. I was wearing a T-shirt, my arms bare. I have been collecting tattoos, travel stamps and emblems for years, but I don't think my grandmother had ever seen the ink on my arms before. I was expecting a reprimand, but instead she smiled. She started talking about her brother, who had been to sea too, a sailor in the merchant navy. She told me how he always had an uneven number of tattoos before a voyage, a promise that he would come back to port for the next. I never knew. In an instant, the gulf I had assumed existed between us, a gulf of my own creation, was bridged. Instead of a woman who would not understand, now I see a woman who did not have all the chances I have had. A working-class woman who understood stability and made sacrifices to create it. A woman who cares deeply for her family. A woman who takes every compliment she is given with a nod and with thanks, in a world where so many of us brush them off. And a woman who beats everyone she plays at dominoes, with a wicked smile. She may not have travelled the world, but she has lived a life of fulfilment and gentle contentment. Now, everywhere I go, I take a part of her with me as a part of myself, in a little gold locket.

My own mother flew a little further from the nest. She was a naturalist from an early age, closely watching the flight of birds and learning the shape of them with pencil, ink and paint. She was the first person in her family to go to university, her curiosity drawing her not to catalogues of taxonomy and a microscope, but to understanding the natural world through art. She pursued her artistic career with a passion and a tenacity that was driven both by a love of creation and by the fact that any fall she made would have been without a parachute. Her family were scared for her and they did not understand her path. She moved to a city made from honey-coloured stones to study, until, later, life took her west. She fell in love with the wild shores of Pembrokeshire in an instant, feeling something in that land, that sea, which called to her. With her savings, money earned through illustration, she bought our storybook cottage, nestled in a ridge of hills that drop to steep cliff, and moved in the very next week. It was a house paid for with paintings, with stories packed into the walls and the hearth. She made her world bigger through her creativity, and she gave me the tools to make mine boundless.

One of my earliest and strongest childhood memories is swimming in the sea with my mother. There is a beach near us where rock meets wild water, where storms swell, where ancient stone is worn to grains by time and tide. The sand there is dark. Granite. The sun was bright. It warmed my skin, and I tried to save the feeling of it even then, a store for the cold that winter would bring.

'Come on.' My mother took my hand. We walked towards the water. 'You need to feel the salt on your skin.'

Together, we stepped in. The waves lulled at our ankles, drawing us deeper. Deeper again. Deep, until the water swept me up. My mother pulled me on to her shoulders as she swam further from the shore. Further than I had ever been before in my early swims, but I wasn't afraid. I clung to her warm body, the last of the heat there staving off my shivers. The water moved around me. Moved through me. I felt the shape of it, the subtleties. I did not know then that this love for the ocean would shape my studies, paintings, words. That the water would take me both from and to loved ones. But the sensation of moving with, moving like, the water stayed with me, and is with me still. I rode back to the shore on my mother's shoulders, to sit on the beach eating summer-sweet nectarines, slice after slice, the taste of the earth. My mother put pens in my hand, an ocean in my heart. She taught me to listen to the world, the strengths and symbols in the quiet moments. I saw her work her way through heartbreak, putting miles underfoot on the coast path and laying careful brushstrokes down on paper. She took me to watch the porpoise feed at the turn of the tide, bewitched me with stories of the waving arms of sunfish. She read poetry to me. Now we write poems to each other, often thousands of miles apart. She told me that no time was wasted if you were happy. She taught me that happiness is not a given, but a practice to be cultivated every day. She taught me to always listen to your own voice, to follow your own intuition. She showed me that you can make a living, build a life, from your passions. She never pushed me, but quietly encouraged, in the hope that I would find my own way.

But, just as young whales are raised not only by their own mother but by a whole community of female whales, so there are other significant women who helped to guide and shape me. Rachel Carson and Dr Sylvia Earle are two of them.

Rachel Carson was an American marine biologist and fisheries scientist. She is most famous for *Silent Spring*, her 1962 book about the catastrophic environmental damage caused by pesticides, which eventually led to the phasing out of the pesticide DDT in the US. But two decades earlier, in 1941, she published her first book, *Under the Sea Wind*. Although it was a critical success, it only gained a wide readership a decade later, when it was reissued following the success of its sequel, *The Sea Around Us*. Through her carefully metered prose, these two books enabled her to bring to life an entire ocean, from complex interactions to individual species, not only for her contemporaries, but for later generations of readers. Her writing was a taper that to this day lights the spark in many young marine biologists. Carson had the ability to transport the landlocked reader from the river, the shore, out into the sea. Perhaps she did not write to transport merely her reader, but also herself. As a woman of her time, Carson actually spent very little time at sea. Her observations came from hours in salt marshes, on the rocky shore, on fishery docks. She marvelled at the phosphorescence on the shore, recording in letters that it was as if there were diamonds and rubies, valuable jewels littered in the surf.

For want of opportunity to observe the phenomena at sea first-hand, she exchanged letters with sailors such as Thor Heyerdahl, and read Darwin's account of his voyage on HMS *Beagle*. You can hear in her words as she describes these 'enviable' voyages her longing to be a part of them herself, out at sea, with only a stretching horizon and no sight of land. Heyerdahl described for her the glint of moonlight on offshore waters and the flying fish that would strew themselves around the deck, the alluring lanterns of squid that lit up the deep as he drifted on his raft *Kon-Tiki* towards Polynesia. In *The Sea Around Us*, Carson talks of depths in fathoms rather than metres, of the discoveries of man instead of human, a gendered world dictated by patriarchy with a heavier weight than it is today. She talks of the giant sperm whale, a large toothed creature which she estimates to be able to dive to a depth of 540 fathoms (3,240 feet, or 987.6 metres), based on a sperm whale that was found dead, entangled with a submarine cable which was raised to the surface off the coast of Colombia in 1932. She talks of when the sea was – to human ears – a silent soundscape, and mentions the ground-breaking discoveries made by the deployment of military hydrophones, which showed that our waters are seas of sound. *The Sea Around Us* suggests that fish, shrimps and porpoise all contribute to the symphony of the sea, with many sounds still to be identified. I wonder what joy she would have experienced if she was able to hear the click, click, click, the single loudest biological sound in the ocean, the rhythm of the sperm whale in the deep.

Carson writes about the destruction caused by human beings to island biodiversity as a result of exploration, exploitation and colonisation. She writes of endemic bird species meeting extinction due to the introduction of rats transported on ships, which eat eggs and young, and destroy nests. She writes of a rising sea level, hypothesising that the edges of the sea will flood ashore until the tideline laps against the Appalachian Mountains. She writes with muted rage, tangled with excitement, about pushing the boundaries of science and exploration that constrained her era, the possibility of discoveries with the advance of technology and the new means of venturing into both surface waters and the depths. What is absent from her vernacular is something that is unavoidable in a book about the sea now, and that is the damage caused by anthropogenic climate change. What would she think of the world of today? It is certain that we have never empirically known more about the sea. But what is also clear is the level of damage we as a species have provoked and intensified. I cannot help but imagine that she would be sharpening her pen, a steadfast voice of warning against a rising tide of harm.

Dr Earle, also American, is an oceanographer, explorer, diver and ocean advocate, sometimes referred to as 'Her Deepness' as a result of her exploratory endeavours below the surface. She is the founder of the organisation Mission Blue, holds a PhD in her own right, and through the course of her career has been awarded twenty-two honorary degrees in recognition of her contribution to the exploration and conservation of the world's oceans. Dr Earle was *National Geographic*'s first female Explorer in Residence, has been named *Time* magazine's first Hero for the Planet, and was the first female chief scientist at the US National Oceanic and Atmospheric Administration. She speaks with conviction and passion about the ocean and the damage we are causing, and for many her words have served as a rallying cry, a call for action on an individual level, within the community, and at a political level, globally.

What has always drawn me to her work is her sound iteration that the health of the world's oceans is the health of *all* of us. That nature does not exist in a vacuum, and the damage we do to nature in turn damages the fabric on which we construct our lives, the fabric of biodiversity. Her own journey started in Florida Keys, where she would spend hours as a child in the seagrass meadows exploring the life she found there, only to see those meadows 'reclaimed' from the sea as they were drained and turned into car parks. When asked in an interview with the Natural History Museum in London exactly what one should do to preserve the oceans, Earle answered that we need to examine where our strengths and talents lie. She said that conservation action can be taken through science, through exploration, through government and policy, but also through art and literature and poetry. That we must all examine where our contribution is strongest, and, once this is identified, we must get to work. She advocates for nature-based solutions, and highlights the fact that we can help the oceans even if we live in the most landlocked reaches, by starting to respect, protect and restore the natural world around us, which in turn will feed into global biodiversity. She talks of the power and influence that is in each and every one of us to affect the world around us, whether positively or detrimentally.

Earle is a pioneer, a scientist who rightly assumed that she should be a part of a male-dominated conversation. In documentary interviews, I have heard her asked why at eighteen years old in the 1950s she was pursuing a career in marine science instead of dating and getting married. When working on her doctorate, she was invited on an expedition to the Indian Ocean to collect data on what marine life we could find in the ocean; the *Mombasa Times* ran the headline 'Sylvia Sails Away with Seventy Men, but She Expects No Problems'. When she saw a flyer at Harvard asking for divers for the Tektite Project to spend two weeks in an underwater habitat in the US Virgin Islands, she applied instantly, not considering that the organisers did not expect to receive female applications. Although she was rejected from Tektite I, when the project was repeated, Earle earned a spot, along with four other women. The documentary *Mission Blue*, depicting some of the life and work of Dr Earle, flashes a newspaper headline that appeared at the time: 'Five Gal Pals Plunge with One Hairdryer'! Although women in science still face many more hurdles than their male counterparts, and their treatment by the press and social media can still be absurd, or worse, much has changed since Earle's early career. She herself has been instrumental in changing the perception of women in science. When faced with frivolous interview questions, Earle used them to turn the conversation around, and to masterfully bring the dialogue back to marine issues. Her interests and range of knowledge are far-reaching, from an understanding of individual species to broad oceanographic processes. As well as the impact she has made in marine studies, her presence in this field has inspired many more women to follow in her wake.

On the quiet nights aboard *Balaena*, when the sails were set and the wind steady, I found myself thinking back to Earle's research trips on the surface of the oceans, and wondering afresh at how intimate and independent an experience of the sea I was able to have. Earle carried out her research aboard large steel motor vessels, where she had no part in steering the boat, or indeed in the day-to-day maintenance. *Balaena*, our small sailing boat turned floating laboratory, was an entirely different set-up. We all took turns helming the boat, setting sails, keeping the boat's log. While all of the acoustic data from the hydrophone was being recorded on to hard drives for later analysis, we had to monitor the electricity supply very carefully. While we were sailing, the recording software continued to draw power, as did the navigational equipment for the boat, and we had no back-up generator or solar power. All of the visual sightings data, along with environmental recordings of sea state and sea surface temperature, were logged by hand, on to waterproof paper, carefully attached to clipboards to prevent the wind snatching it away. We cooked, we navigated, we hoisted and lowered sails, reefed when we needed to, all the while conducting observations. We worked through each and every night. If something broke, there was nobody else to fix it save for the five of us on the boat. It was hard work, but it felt good to be involved at every level. Neither way of researching is better or worse, just a different experience.

The voyage was coming to an end, and I felt a mixture of emotions. I was dreaming of a long, hot shower, of finally peeling off the layers of thermals that seemed at that stage to be more firmly attached to me than my own skin. I was eagerly anticipating a full night's sleep, yet I knew I would miss the crew, the closeness, the idiosyncrasies that had developed during the voyage. And I would miss the companionship of the whales we could hear in the depths, engaged in a conversation we cannot understand. I wanted to smell gorse and heather, to dig my fingers into the soil and absorb the smell of the earth, yet I knew I would miss the silky hours sailing under moonlight, when you felt like the world was yours and yours alone, the ocean whispering stories. Summer was coming to an end, and the autumn would bring changes. I was for the first time to enrol in full-time study. I wondered how it would feel to sit in a lecture hall full of students, seventeen- and eighteen-year-olds trying to navigate life away from home, likely for the first time, as we learned about whales – instead of sharing space with them, out on the sea. I wondered if I would still have the time to watch every sunrise, every sunset. I knew I needed to learn in this more structured way, but I was worried I would feel out of place on land again. As we continued to sail the track, I started to find myself willing the miles to stretch, the hours to slow. Besides, our mission felt unfinished.

On that voyage, we had been following a 1,000-metre-depth contour line, where the seabed dropped away a kilometre underneath us, approximately 300 nautical miles off the Newfoundland coast. The whole area was considered a suitable habitat for northern bottlenose whales, a member of the 'beaked whale' family, but they had never been recorded there. For weeks, we had been sailing, towing the hydrophone and listening every fifteen minutes round the clock, day and night, to see if we could detect their vocalisations. Dr Feyrer told me that the analysis of the data in the lab showed that we had actually encountered the whales, but either we had failed to identify their calls, or they hadn't coincided with our listening periods. One thing for certain was that we hadn't seen them at the surface. With just a few days left, we headed to the Gully, a submarine canyon off the coast of Nova Scotia. Previous research had identified a population of northern bottlenose whales there that have shown high site fidelity since they were first studied in the late 1980s. After tracing these elusive creatures through the Atlantic all summer long, I was longing to catch a glimpse of them. As we sailed south, the days were growing warmer and warmer, and I, more and more conscious of the distinctly human smell I must be emanating, was reluctant to shed my layers. Those final sunsets of the voyage were so vivid, the colours stained in my mind as the water was painted blush and gold in illuminating strokes at the dying of the light. We would fix our eyes on the horizon while we ate dinner, waiting for the fabled green flash, the brief burst of colour that occurs on occasion at the moment of sunset. Each meal was increasingly just a variation on the small amount of fresh produce that was still holding out – you would be surprised how many concoctions you can sneak celery into.

It was around midday when someone aboard called out, 'Beaked whales!' I don't remember who saw them first, only my incredible excitement. Beaked whales, classified under the family *Ziphiidae*, are treasures to find. They are the most under-studied family of whales, known to be the deepest divers, venturing further into the depths than even the sperm whale. They are typically hard to discover in offshore habitats, and can be extremely elusive and unapproachable at the surface, often diving before you even know they are there. Most of my encounters with beaked whales have been entirely unexpected, when they have breached out of the water with a joyful vigour on days when it has been so rough that it seemed hardly worth conducting visual observations at all, only to disappear again into a swirling cacophony of wind and wave. The northern bottlenose are the exception, as they can be gregarious and are known to approach boats. Although this makes them slightly easier to study, they are also more vulnerable to whaling. I squirrelled my way up the mast to get a better look. At first, I saw a light grey shadow moving under the surface, over half the boat's length in size. My view was lurching, from the movement of my platform and from excitement, as I leaned out from the crow's nest. Holding on to a halyard to keep my perch, I watched as the animal surfaced, its head breaking the water first. Pale grey in colour, skin glistening in the sun. It had a beak of a nose, akin to a dolphin, and a bulging melon of a head, the jut of the forehead angled at almost ninety degrees to the beak. Then came the blow – quiet, a light puff – followed by the elegant curve of its back to the dorsal fin. The tail, sickle-shaped, stayed in the water as it continued to swim with ease ahead of the boat. There were three more animals converging in front of us, milling together before turning to approach us. We watched the whales and the whales watched us back, a moment of mammalian recognition.

And then the work began. The rest of the day was a blur as we helped Dr Feyrer to gather as much data as possible: ID shots of the animals and samples for genetic analysis, to be put into liquid nitrogen immediately in order to be transported back to a lab. There is always a fine balance between gathering data on the animals and the level of disturbance, so Dr Feyrer was filming the entirety of our interactions with them to help maintain a minimal acceptable level, and we constantly monitored their behaviour, looking for signs indicating disturbance, such a tail slapping.

As I lay in my bunk in *Balaena* for the last time, I was aware of how much work lay ahead, of how much time I needed to spend at sea to develop the casual fluency of seamanship I admired in more seasoned sailors, but I felt my course was set. It was time to return to shore, to gain a fresh perspective on the oceans. My time on *Balaena* was over, but for Dr Feyrer this was just the beginning: during four summers spent afloat, she would finally locate and study the population of northern bottlenose whales in the waters north-east of Newfoundland which had been reported anecdotally by fishermen but which had eluded us. The following year, during her fieldwork, she went back to the same location where we had had the acoustic detections. This time, she had visual sightings of the whales and was able to take samples. In 2017, she returned again, and concentrated her survey in this area, concluding from the continued presence of the animals, both adult and calf, that it was an important habitat for the northern bottlenose whales. Her work will contribute towards future protection of the species in this area.

I recently spoke to Dr Feyrer, reminiscing over the shared experiences we had on *Balaena*, and talking through her thesis and publications. She mentioned that she had dedicated her thesis to her daughter Ione, and spoke of the shared responsibility and family support that had enabled her to conduct field studies while also bringing up her child. I thought again of the clans of female sperm whales, of the community effort involved in raising calves. And I thought of the support I had felt from my own family, the line between my grandmother, my mother and me, of the inspiration I had found in the women scientists who had gone before me. Aside from Dr Feyrer's academic achievements, what struck me from the conversation and continues to resonate is how she talked about the time spent with the whales. She talks of her delight in the element of the unknown, and how rare it is to encounter this in twenty-first-century life. When you set out on a voyage to study wild animals, particularly marine mammals in their natural habitat, you never really know what you are going to get, the interactions that will follow. She spoke of the endorphin rush we experience as humans when in the presence of large mammals, and the deep laughter and sheer joy that accompany time spent with whales. It is a privilege to be, for a short time, in their element.

3

Human

There was sunlight streaming through the windscreen of the camper van, dust motes dancing in the rays. The gentle rhythm of a rolling swell filled my ears, setting the tone for a blue-sky day. There was warmth in the air, the touch of spring sun breaking beyond the resounding depth of winter. There was celebration in that light. Renewal. My collie dog sighed and stretched as I slipped out of bed. There was no need to kindle a fire in my tiny log burner that morning, so I started to whisk batter for pancakes. Hot coffee, blueberries, their skins bursting to soft insides, maple syrup poured over everything, the sweetness a mirror of the sunshine. I pushed open the sliding door with a practised little kick where it always got stuck on the rail, and sat on the step, the collie beside me.

Our view was the beach. The sand was divided into sections with rotting wooden groynes to limit the movement of the sediment, cross shore drift. They jutted out of the sand, then out of the water, until little by little they were swallowed by the North Sea. Blue sky into blue sea, two orange tankers anchored on the horizon, a splash of colour. Our backs were to the ugly complex that makes up the modern part of the Aberdeen waterfront: apartment blocks, a drive-through, pound shops and a supermarket. Looking seawards, you could almost forget that they were there. We lived here, my collie and I, in a camper van by the beach, in north-east Scotland. I was studying at the university – old stone, stained-glass windows, ivy creeping – learning the sea by numbers, statistics, studying the anatomy of the creatures that fascinated me.

When the voyage on *Balaena* came to an end, I had struggled to adapt to this new environment, the shift from days and nights dictated by the weather to a timetabled schedule of lectures and labs. I often felt tired and crowded when surrounded by people, and life was full to the edges – trying to find my place in academia and fund my course. There were times when I had to run straight from dissecting fish in the lab to the yoga class I was teaching in the evenings, hoping I did not have fish scales lodged anywhere, and that I hadn't carried the smell out of the lab with me. But it was also a relief to be in one place, focused on one thing. Living out here on the fringes of the city, I had my space, and there was freedom to the way I could spend my time. Gradually, I began to find my way.

My mornings would start by or in the ocean, a swim or surf, before cycling to university to study. On Fridays, I would take off with the dog to explore the Cairngorms, or the Scottish islands, discovering a landscape that was simultaneously new and wonderfully familiar. The previous weekend, we'd gone to Skye. We were by the sea, before the scar that is the Cuillin Ridge reaching into the sky like the dark teeth of the earth, when I stumbled across a minke whale, dead and washed ashore. This creature was in an entirely different stage of decomposition from the pilot whale I had seen when I was nine. Flesh still clung to its giant vertebrae, stringy and sinuous, hanging off the spinous processes. Most of its throat had gone, and its stomach was split open. The fresh breeze carried the worst of the smell away, so I could bear to get close, to discern the majestic animal in the parts that formed it. Over the coming months, the rest of the flesh would fall from its bones, picked over by birds and detritivores until only the skeleton remained, a white sculpture on that isolated beach. Perhaps it would be washed back out to sea, to rest once more in a world of water. Perhaps I would follow, go back to sea – but not yet. I was still buoyed by my time with the whales in the North Atlantic: the shower of sperm-whale spray in the night; the dark, dominant fins of orca as they cut through the water; and the small curved dorsals of Cuvier's beaked whales as they surfaced against a rose-tinged sea.

Every night, I would light the tiny log burner in the van, close the doors and curl up with my collie, her soft snores as soothing as the sound of the water. The van was the first space that ever really felt like my own beyond the walls of the cottage, perhaps because I had taken so many pieces of it with me, physically as well as mentally. My kitchen counter and cupboards were modified from an old Welsh dresser that used to stand in our kitchen, my table the lid of an old pine chest where we used to keep our winter jumpers in the hallway. All of my worldly belongings at the time fit into that small space: paints in one drawer, a short shelf of books, surfboards tied to the roof. I didn't have much, but I didn't need much. The tiny wood burner would only fit logs that had been split thrice, but they kept us warm all through the Scottish winter, took the chill away after morning dips and made the nights feel ensconced.

That morning, after breakfast, I climbed up on to the roof of the van. One foot on the passenger seat, a pull up with a swing. Careful not to tread on the solar panel that provided my electricity, I untied a surfboard and lowered it to the ground. I tugged myself into my wetsuit, which was still unpleasantly damp and cold from the day before, and squeezed my feet into neoprene boots. I left off the gloves and the hood for the first time that year, enjoying the feel of the sun on my face and the sound of the water without the muffle of the hood. I headed to the south end of the beach, right where the North Pier hooked around for the entrance of the port. I'd walked here with my collie the first day I got to the city, astonished to see the fluke of a humpback whale, dappled grey and white, which had taken up residence around the harbour apparently undeterred by the shipping traffic. It had felt like an omen, a sign that I was in the right place. Although the humpback left just over a fortnight ago, more often than not I would see bottlenose dolphins (*Tursiops truncatus*) around the harbour entrance, their fins and backs breaking the water, or large shadows flitting just under the surface.

It was strange to see these creatures that I associated with wilder waters come so close to the high-rises that make up the beachfront: the juxtaposition between the fluidity and luminosity of their element, and the harsh lines of gleaming chrome and glass throwing out a weak imitation of the light on the water. Aberdeen Harbour itself is exceptionally busy. The port is the largest hub for the support and supply for North Sea oil and gas in the UK and Europe. Tankers constantly rumble in and out of the navigational channel, as well as smaller boats carrying crew transfers and delivering supplies. Aberdeen also has a fishing fleet and is a departure point for NorthLink ferries, connecting Orkney and Shetland to the mainland. The bottlenose here are considered resident, one of three coastal populations in the UK and Ireland. The others are found in Cardigan Bay, just north of where I grew up in Pembrokeshire, and on the west coast of Ireland. For the population in north-east Scotland, the Moray Firth is the northern extremity, stretching to the Firths of Tay and Forth to the south of Aberdeen. The Moray Firth has been named a Special Area of Conservation (SAC) as the result of a long-term photographic identification study there, affording the habitat of the bottlenose some protection against the expansion of coastal and offshore development, but the presence of the species at the mouth of Aberdeen Harbour suggests that the conservation area may not be large enough.

Coastal bottlenose dolphins live where they can find prey, and Aberdeen Harbour stands at the mouth of the River Dee, which has long been rich in salmon. One study of the bottlenose in Aberdeen saw the dolphins travelling along the deep navigational channel in the wake of the large ships entering the port, feeding on fish churned up by the ships' propellers. Although it seems that the dolphins have learned to live with the boat traffic, it does cause a change in their surfacing patterns and sometimes their breathing can become wheezy. Some have learned to temporarily avoid certain busy areas. It is likely that many of these dolphins experience prolonged stress due to the physical presence of the boats, increased noise levels in the water, and disruption to time spent feeding or socialising; over the long term it may be possible to confirm this by looking at the rate of reproduction amongst these dolphins and the health of any calves. But it is hard enough for us as humans to pinpoint or recognise stress in our own lives, so easy to get caught up in it and consider it as a norm, let alone to be able to recognise and study stress in a wild aquatic mammal without invasive methods that cause more stress.

I walked to the sea in the sunshine, threw my board in and started to paddle. The water had a longer memory than the air, the sting of cold harsh on my bare hands. The surface was glass-like, the swell small. There were a few people in the water already, bobbing like seals in their thick wetsuits. After a friendly nod, we kept our distance, each content to have a piece of the world to ourselves that day. The sky was reflected in the sea, the heavens trapped in the water, made other-worldly with the slight roll of the surface. I looked over my shoulder to see a small set rolling in, picked my wave and started to paddle. I started to speed as the water rushed up to greet me, and, with a moment of exhilaration, I took off, flying steadily over the surface. I don't know how it looked, but it felt graceful, flowing. It felt right.

Ahead of me, I saw the grey shadow, as the fin of a dolphin, a large male, broke the surface, riding the wave with me. In that single perfect moment, we were just two mammals, sharing a ride on a wave, rolling towards the shore, for no reason beyond a simple, unadulterated pleasure in being alive. I had a Cell Biology assessment that morning, a fifth of my mark for that course, but I could not tear myself away from the sea. I knew I had done enough to pass the course without it, but I would have stayed anyway, found some way to make it up later. The dolphin pulled back as we approached the shallows, and I too dropped off the back of the wave to paddle out again, far slower than my slick companion. I caught wave after wave, until my arms were jelly, and, although I rode the rest alone, I tumbled ashore smiling, grateful, humbled. In my work, I am always aware of the danger of anthropomorphising our human intrusions into the world of cetaceans. But there was something so serendipitous about this encounter with a wild creature – the sense that it was sharing space with me of its own volition, me riding on the surface, the dolphin skimming underneath.

The sea is fickle, drastically changeable in the space between one breath and the next. The fairest day can become foul with the turn in the tide, a shift in the wind or the marching advance of a weather front. I don't know if I loved the sea for this because I saw something of myself, a Narcissian reflection in the fluctuating surface waters; or perhaps I merely learned my way of being from the waters that raised me. Just a month after that perfect day, eating pancakes in the sun and surfing with dolphins, everything felt different. It was hot and close, the air so thick you could chew on it, pathetic fallacy in cloud and heat, a storm waiting to break. My headache was building, the pressure against my forehead growing as the barometer took a swooping drop. Change was in the air, in the water, as a confused swell hit the beach. And although I didn't yet know what was shifting within me, I recognised the feeling of incipient change. It was something I had known from a young age: that the foundations on which you base your life can crumble into uncertainty before something new is built.

There is a Celtic folk tale I had been told since I was a child. My mother, my brother and I would walk from the cottage, up the stone-lined lane to the brow of the hill. I always wanted to walk faster than them, to be the first to the top, to see the sea laid out before me. We would take a path to the left, Carn Llidi in front of us, the highest piece of land for miles, its silhouette stretching all the way down to the sea in the evening light. The path picked its way through gorse, and the passage had to be negotiated with the ponies. After gorse came bracken and fern, and, if you looked closely enough at the end of spring, the tiny purple pinnacles of marsh orchids. The descent took us to our destination, a ruined village, Maes y Mynydd, meaning the mountain field, place by the mountain. The settlement is dated to 1829, a Quaker village, next to the sea. The inhabitants, like many Quakers in Pembrokeshire, were driven off the land and never returned. There is a rumour that to get them to leave, the landowner poisoned the village well, turning the water against them. The village has been left to time, quietly crumbling, stones falling, lintels rotting in the salt. I remember when the last chimney *fawr*, the large fireplace that would have been the heart of the home, fell down, a strange finality in a place that has been empty for centuries. We would walk carefully between the walls, before finding a place out of the wind. My mother told us her own stories of the inhabitants, the ideas that had captured her imagination when she first walked these paths – not of Quakers, but of a villager and a selkie woman. Selkies were not her own creation. The story always has the same bones wherever it appears, but the details vary with time and telling. Selkie, seal wife, woman of the sea. A stolen bride, or a good-luck charm, sometimes both. My mother had her version of the tale, just as I have mine.

This one is the story of a fisherman and a selkie woman. The fisherman was a man of few words, with deep eyes. He lived in a village, tucked away at the edge of the world. He lived a slow and simple life. He fished, tended his garden, kept bees in hives, goats for milking. At night he sat by his fire and sang to himself. He had a lovely voice, a voice they said could charm the very fish from the water.

One night, as the fire burned low in the hearth, the wind and moonlight started to beckon, whispering through the windows, under the door, drawing him from his bed and to the shore. He walked out into the night, and under a bright moon traversed the precipitous cliff path. Sitting on the pebbles at the edge of the sea, he started to sing, his voice clear in the soft night.

Dark eyes in the water, a ripple, as a seal swam to the shallows. From slick glide to a heave, she hauled herself over the pebbles and up the beach. As she made her awkward transition from sea to land, her pelt seemed to be straining, turning, elongating. The water-slicked fur was falling away to reveal soft skin. Fins became arms, as for a moment she pulled herself along. Her tail, designed to twirl her way through the brine, now twisted and grew, contorting until she had legs on which to walk. The fisherman believed himself to be asleep, so in the carefree reverie of a dream he grew bold, his voice louder as he stepped towards her. He expected this seal, this woman, this selkie to perhaps flinch at his presence, or shy away, startled, as she realised that she had been summoned ashore by his song, like the fish to his nets.

As the heather of his eyes met the storm-grey of hers, a wicked smile lilted her mouth, and he realised that he was not in fact the hunter but the snared. She held his gaze, reached out a sure hand. In the space of a breath he took it, and they began to dance. Theirs was a dance as familiar as the sea meeting the shore. They danced on until the mists of dawn began to fall and the sun brightened the east.

As day broke, he was sure it would break the dream, that the selkie woman would vanish and he would wake in his bed. But that was not so. He missed a step in his confusion, and the dance faltered. Next to him lay her sealskin, mottled silver, inky dark. He picked it up, and turned to carry it to the woman. Its slip and slick were so far from the rough wool he was used to feeling under his fingers. Moonlight, velvet, secret, water.

'You could take it, you know. Keep it safe. But only for a time. If you like.'

He knew the stories, he knew how this worked. If a man stole the slipped skin of a selkie while she changed her matter, he could keep her for his own on the land, a treasure stolen from the sea. But he was no thief, had no wish to capture. And here she was, an offer made, a bargain, to stay. If only for a time.

Together they climbed back up to the village. She lay her sealskin in a chest at the end of the bed. For that day and night he neglected his nets, his garden. The walls of the cottage were their only world, and how they spent their time, only they truly knew. There was talk in the village. Those who ventured close to the cottage heard him playing his fiddle, heard him singing. Alongside the familiar music flying from the fiddle strings, there was a new song, a new voice. Although they strained to listen, the words were in no language their minds knew. It was strong and deep, filled with something so beautiful and complex that it was both impossible to understand and impossible to forget. For all the might in that voice, it sounded lonely, a haunting echo along with a coaxing lullaby.

The next morning, they ventured out into a bright day. The villagers whispered and watched from the shadows. Perhaps it was her voice they had heard. At first the rumours flew, but then they faded, as she settled into life with the fisherman, accepted by all as his wife. In time, nobody remembered her arrival; to them, she had always been there. She lived with the fisherman, and they made their life together.

When their child arrived, they joked about whether it would have fins or feet. She was a baby girl, perfect, with the other-worldly beauty of her mother, and the heather eyes of her father. She had a piece of both of them in her voice, the sound the sea makes where it meets the land. She was lithe and quick in the water, but had a place on land too. Her mother never tried to conceal who she was to the child, where she had come from. She opened the chest, showed her the sealskin. The girl marvelled at the colours in it, she felt its weight, its slick. On nights when both her parents went out to fish, she would wrap herself in it, imagine herself beneath the waves. When she walked out of the village and sat on the shore, she would call to any seals she saw, wondering if they were kin.

'Mother, will I be a selkie?' she asked one night by the fire.

Her mother looked up from her work.

'Your life is your own, my child. Your path is yours to choose, but in choosing you must walk it. All I can tell you is that, in time, you will know your nature. And to go against one's nature is a difficult and dangerous thing.'

The words seemed to carry with them a heavy weight, and a tear fell from her mother's eyes.

'Mother, why do you cry?' the girl asked, afraid to hear the answer.

'The tide is changing, my child. The music slowing. Perhaps, at last, the dance is coming to a close.'

'Does this mean . . .'

'Don't worry. Don't you worry one bit. It is my time on land that is coming to an end. Never my time with you.'

That night the fisherman woke. His bed was warm beside him, and yet it was empty. He rose and saw that the chest at the end of the bed was open. Open and empty. He walked out of the house and there she was on the beach, skin bared to the sky, staring out at the water. She turned as he approached, and once again, the storm-grey of her eyes met the heather of his, as she reached out a hand and they began to dance. The steps were different, no longer the hunter charming its prey, but more resonant, familiar, happy, sad, lost, found. He wanted to speak, to ask her to stay, to ask that the night might never end. But end it must, as well he knew. As the light started to rise in the east, they found themselves next to the water, her pelt laid out on the rocks.

'My time here is over. My bones feel too heavy, and I know in my blood I must go. I won't come to land again. But when you are out on the water, sing to me, and know that I hear you.'

He knew. He had always known. In loving a wild thing, you have to be able to let it go.

'I will. I will sing every night.'

After his last words to her, she began to walk into the shallows. Her skin rippled, stretched and grew as she walked deeper, until, at last, she dipped her head. When she surfaced again, it was the slick head of a seal that broke the surface, and he watched as she swam further and further out, hoping she would look back, knowing she wouldn't. She was seal now, his wife gone.

The fisherman went back to the cottage and held his daughter close. For all that he felt an emptiness, knowing that his selkie wife would never return, there was also a sense of finality: the promise she had made, to only stay for a time, had been kept, the circle had been closed at last. In her wake she had left behind that which was most precious to him in the world: their daughter. He did not need to explain to her; she already knew. All day he held her, and that night it was she who sang him to sleep by the fireside.

In time, he went back to his life in the village, as steady as the roots of a tree. His wife faded from memory, and the people of the village would have forgotten all about the selkie woman, save for her daughter. Dark-skinned, dark-haired, a storm in her eyes, but a laugh that spoke of a summer's day. They remembered that, at one point, her mother had been there with her, just as beautiful as she.

What became of the girl is another tale for a different time, but the story of the fisherman and the selkie woman became a part of my story in its own way. There was less magic and mythology, more shouting, but my parents' marriage broke down when I was too young to understand, so young that now I hardly remember them together. What I do remember well is the hurt and confusion that followed, the tears and stress when shuttling back and forth between houses until life settled again. I was the first child in my school whose parents had separated, and the split felt loud in a quiet coastal town. Other children were curious about a situation that was different from the norm. One day when my mother was dropping me off at school, a child asked why she wasn't wearing her wedding ring.

'Oh, that,' she said.

She had hurled it off a cliff, into the Irish Sea.

'It's with the selkies now.'

I would trace the clifftops, staring at the seals, wondering if they could be selkies. I wished that when I got home, my own story would be just that, a story, rather than the tumultuous reality. Later in life, the selkie story would resonate with my feelings of being torn between land and sea, but as a child I associated this divide between elements with having to move back and forth between houses. The weeks I spent away from the cottage, in a flat in town, away from my cats, my attic room, the storms, the lighthouse, were the weeks I would feel confused, adrift – the beam of light replaced by street lamps that offered no comfort to my late-night thoughts. So instead I would turn on the Shipping Forecast, allowing its repetitive tones to take me to Dogger, Fair Isle, Fastnet. I loved my father. I wanted to spend time with him. But I didn't want to have to leave my home, my cottage on the coast. As much as I hated the situation, I hated the idea of change more.

All of those old feelings from childhood now began washing

over me again as I felt the tide turning within my relationship with my partner. I had felt that he was my best friend. We had grown up together, worked together, sailed together. We had made our home together in a camper van that used to be an old fruit-and-veg delivery van. We had spent large swathes of time apart, through respective sailing trips and my studies, but we had loved each other like very young adults do, and really given each other space to grow. This familiarity, this companionship, was a foundation on which I had built my life. Now it was falling apart, and I was furiously angry, with all the passion of my twenty-two-year-old self. Angry at him, angry at myself. In reality, all that had happened was that two people had got caught up in their assumptions of how they expected each other to be, and could no longer see what actually was. This young heart-break would have been just that – a coming of age – but what happened next, when I was already feeling slightly untethered, culminated in setting me adrift.

As ever, I sought solace in the water, heading into the surf that was breaking with the storm. The waves were steep, unforgiving, but so was I. My arms ached as I paddled, heart pounding, breath hard. But as soon as I made it through the lines of breaking waves, my adrenalin died, my anger and frustration turned to confusion. Grey walls of water were stacking up around me. I turned my head from side to side, whipping it round, trying to read the water. The surface gave nothing away, every sound just seemed like roaring. I paddled frantically, afraid, tried to take off. For a minute I was going, carried along with the wave, and then, without intending to, I pulled back. I didn't feel in control of my own body. It was too late, and the wave took me with it anyway, tumbling head over heels as the crest broke. Surfacing, gasping for air, eyes and nose streaming, I grabbed for my board and tried again. Again. Again. I couldn't do it, I couldn't get it together. I was never in the right place. Up was down, down was up as I was tumbled over and over. The water no longer held me, it threw me, a cat playing with a mouse. I did not know this water. I did not know this self. My pulse was racing, my breath fast, shallow, panicked. My chest felt like it was being pressed inwards. I knew I needed to get to the shore, and started to paddle. A large wave caught me from behind. I hadn't heard it coming, the noise inside my own head suddenly too loud for me to hear the water. I couldn't even fathom trying to ride it as the water crashed. It jarred me down hard, into my board, and held me there. My eyes stung with tears. I was burning. Lungs. Back. Ankle. I was rolled over and over until the water shallowed and I was unceremoniously dumped on the shore.

Somehow, my board made it through in one piece, but as I tried to stand to walk up the beach, my ears ringing, my ankle felt sickeningly loose. I sat, my first thought to pull my neoprene boot off quickly – I'd just bought them, and didn't want to replace them if my ankle swelled and I had to cut them off. It seemed vitally important not to waste the boot. Just take off the boot. Take off the boot. Take off the boot and everything will be OK. Inch by agonising inch, I peeled off the neoprene, my ankle ballooning quickly underneath. It looked like the puffball mushrooms I used to collect with my mother, which she would fry in butter for my breakfast. Boot off, I hobbled up the beach and over the grassy ridge to the van. There was no way I would be able to climb up to the roof right now, so I shoved my board underneath between the wheels. If someone wanted to steal it, it was theirs. I threw my wetsuit with it, chasing the salt water away as I put on my warm layers. A shot of rum burned down as my pulse started to settle. I couldn't drive to the hospital on that ankle anyway, so I climbed into bed, and pulled the collie close beside me, my face buried in her fur. I pulled the duvet over my head and tried not to think about what had just happened.

The storm passed overnight, driftwood and seaweed tossed ashore along with the inevitable plastic that littered the tideline. If the wind rocked the van, I didn't feel it, didn't hear it. There was just a sterile kind of silence. A void in my head. I tested my ankle, which was still disturbingly loose, mottled bruising starting to come through. My back felt stiff, awkward, like two pieces of a jigsaw puzzle that look right at first glance, but won't quite fit together comfortably. The back of my left leg stung all the way down, as if I had fallen in a nettle patch, or had the long tentacles of a lion's-mane jellyfish wrapped around it, culminating in a hot flame at my Achilles tendon. It was as if a string inside me was being pulled far too tight. And emotionally, too, I had reached the point where I could go no further. It felt like the current that had swept me along all my life had stopped.

As soon as my ankle could take it, I taped it up, packed up, and started the long drive home to Wales, to the cottage by the sea. Term had ended, and I had three months to get myself together. Pembrokeshire seemed like the best place to start, back at my beginning. I've wondered ever since about the relationship between physical, mental and emotional health. As I dragged myself home, the pain was sharp, but I was almost glad to have this physical sensation; it felt easier – something I could see to account for how I was feeling.

My ankle healed quickly, but the pain in my back, the stinging in my leg continued: they felt deeper, internal. As soon as I was on my feet, I was pacing the coast path, hoping the motion would rock my head to stillness. I walked mile after mile, and swam whenever I could, in rivers and in the sea. I went to physiotherapy and saw a chiropractor. I began to sail again, watching moonset and sunrise off Lundy Island, the sun-blushed surface of the dawn sea broken by the exhilarated splash of common dolphin. An acupuncturist carefully needled my back and legs. She was the first person to ask how I felt, beyond the symptoms of my pain. And I realised that day by day, things were feeling easier. As I regained mobility, I found myself laughing with my friends again, whoops and roars as we launched ourselves off cliff ledges and into the sea, like the guillemot jumplings. I scrambled mountain ridges. And later in the summer, I travelled. From the white sandy beaches of the Hebrides to the dusky purple clouds that pooled around Table Mountain in South Africa. For a month, I practically lived at a yoga studio by the sea in Cape Town, both teaching and taking classes. I skated. I climbed. I watched right whales breaking the surface of the Atlantic in the shadow of the mountain. I filled my heart with wonder and felt steady on my feet again. The problem was my back.

Although I felt better in myself, the pain in my back was getting worse. Both my legs stung now, and although I hardly noticed at first, my left leg felt slightly numb, and at times would tingle. Sometimes I would have to mask a scream when I went from sitting to standing, or if I bent down to pick something up from the floor, as a sharp, electric pain shot through me, seeing white behind my eyes. Lying down became increasingly uncomfortable, until the point where I would barely sleep more than two or three hours a night. After months of ineffectual treatment, both my chiropractor and physiotherapist suggested I see a doctor, but the doctor dismissed me without an examination, saying that I was too physically able to possibly have a spinal injury, and saw no need for an MRI. She said that everyone experienced back pain at one time or another. And I tried to believe her. But, deep down, I knew something was badly wrong.

It happened in my second year of university. I had transferred from Aberdeen to Plymouth, the length of the country. The course at Plymouth was more focused on fieldwork and the kind of self-led projects I loved, but, academics aside, I was also glad to be closer to home. One weekend, in early November, I was sailing on the Carrick Roads in Cornwall, a beautiful, tree-lined estuarine stretch of the River Fal as it flows out into the sea, a transitional place where fresh water becomes brackish becomes salt. It feels as though you travel back in time there, encountering the most wonderful selection of boats, built at the turn of the twentieth century – old wooden cutters, smacks and luggers – all looked after with devotion, kept bright with paint and sailed often. On the waterway, there are sailmakers, boatbuilders, welders – anyone and anything you could possibly need to keep a boat sailing. On the River Fal, there is a fantastic female boatbuilder called Amy, who once fashioned me a new tiller to replace a rotten one in her workshop almost overnight, so I could leave on a sailing trip before a heavy storm set in. It is a very singular place, but on that particular November weekend, I could not focus on my surroundings, too distracted by what was going on with my own body.

We anchored in the bay at St Mawes, and I went to step down from the deck of the yacht into a dinghy to journey ashore. The step seemed never-ending, my leg outstretched, hands clasping wire shrouds. But when I looked down, there my foot was, already on the base boards of the dinghy. I couldn't feel it at all. It was the most terrifying moment. I had grown as used to the pain as I could. That sharp white feeling, the burning that stole the sleep from my nights and coloured my days. But this. This was loss. It was as if my foot was gone, or my brain did not know that it was mine. I reached out with my hand. I could feel my foot under my fingers, but it was as if I was touching someone else. The nausea was instant, my stomach lurching. Carefully, with shaking hands, I lowered myself into the dinghy. I didn't tell anyone. I didn't know what to say. The doctor's dismissal had left me wondering if I could trust my interpretation of pain. But now it felt like the worst thing, the thing I had secretly feared, was happening. Over the previous months, I had been able to help my mind, but that had only been possible because of the strength of my body. And now my body had quit. Sitting on a stone wall, watching the boats gently bobbing at anchor in front of us, a friend was talking to me, asking me about my time in South Africa. I could barely comprehend what she was saying, and must have appeared rude in my silence. I was quiet and scared, overwhelmed by a feeling of fragility.

I didn't go to the doctor straight away. First, I moved into a house – living in my camper van was no longer a possibility as I was struggling to walk, let alone drive. I was nervous that nobody would take me seriously, but a friend persuaded me that I had to get someone to listen, so I found myself once again at the doctor's surgery. Grey and beige, muted, sterile. It was a different doctor, who wore a checked shirt – bright, primary blue and red checks shot through with a seam of yellow, a synthetic imitation of the colours of my childhood landscape. Tan suede shoes, scuffed on one toe. He was young, not much older than me. He had kind eyes. He took one look at me and let out a breath. I was half sitting, half standing, neither comfortable any more, an awkward in-between. My jaw perpetually tense, voice shaky. He understood straight away and his voice was so sympathetic that I nearly cried. I was always on the edge of tears, but somehow they never fell. He gave me a referral for an emergency MRI, with the warning that emergencies in rural areas with an underfunded NHS had a six-week waiting list.

He gave me a pack of pills. An antidepressant, to be taken in low doses as a suppressant for nerve pain. At first they made my head spin, made the world blur with any quick movements of my head, and, embarrassingly, my words slurred slightly. But that night when I curled up in my bed – a new bed in a new room – and gazed out over stacked rooftops towards the sea, I fell into my first good sleep in months. I couldn't believe it when I woke up. The pills had worked. Even if they made my head spin, they worked. That week, I studied as much as I could, trying to catch up with the work that was already underway, as we approached the end of the term. I read and read, tapped out code – I was learning new methods of analysing data – drew the anatomy of different fishes. But the pain, though suppressed at times, was still there.

Pain is a curious companion. It is a lens through which a perspective on life can be sharpened, giving you a heightened sense of gratitude for those times before it dogged your steps. It can give you a determination, focusing your energy on what you will do, what you will achieve, when the pain subsides. But I have found that we can only tolerate each other for so long. Within less than a week, I was back at the doctor's. I had tried everything I could to ease the pain, from physiotherapy to yoga, the chiropractor, cutting out sugar from my diet, boosting my greens, putting turmeric in everything, acupuncture, and anyone and everyone's cousin's uncle's sworn remedy for back trouble. Now I had found a doctor who took my pain seriously, I was ready to listen to him, to whatever he told me. The dosage of my medication was increased, and codeine added on top, with a promise that it wouldn't be too long before my MRI. This new combination of drugs was heady. Heady and tiring. I slept a lot, but would wake feeling barely rested, my back screaming, reaching for another pill before anything else. I assumed this overwhelming tiredness was a side effect of the medication; while I am sure it did contribute, I now treat my back with very different methods that are opioid-free and I still experience fatigue. It is a symptom of the nervous system being in distress, and the body attempting to manage pain.

A fog had settled in mentally, thicker than any I had seen on the North Atlantic, and my memory would falter; simple things like finding the right items in the supermarket seemed confusing. I felt vulnerable. And yet that week I sat an ecology assessment and somehow, a rabbit out of a hat, I pulled some of my highest-ever results. I had been unable to attend most of the lectures, but, by a stroke of luck, I had read the required textbook the summer before. The old information seemed lodged in my brain, even while I struggled with the new stuff. Sitting to take the test itself was hard; the weight of gravity seemed enormous on my damaged spine, and knowing I couldn't get up and leave in the middle of it filled me with anxiety. It seemed like panic was my constant companion now, waiting for any situation that felt in any way claustrophobic.

My walk was slow, with a slight limp as I tried to keep up, my friends' regular gait always half a step in front of mine. That half step seemed like a mile. It felt as if I was watching everyone else walk on with their lives, while I fell behind. I tried to walk down to the water as much as I could. There were good days where I would make it, and stare out over the winter sea, and I would remember my why. Why I was still studying. Why I got out of bed each day. There were bad days too, when the walk was too far, or too depressingly difficult. I felt like I was being given more and more to deal with, a tightrope set higher and higher, and yet all my methods of finding balance – running, swimming, yoga, walking – all of my safety nets were being pulled away. The one thing I did still have were pens for my hands, paint, brushes. Every night, I poured out all the things I could never find the words to say out loud, pages and pages written to myself, pep talks, the things I was grateful for. I couldn't sail any more – I could hardly stand on dry land, let alone on the swell of the sea. I couldn't venture offshore, into the world of the whale, so, in my little room, I painted it, the seas, the stars, the skies, the spray.

I painted humpbacks, with their graceful fins and grooved throats. I painted porpoise. I painted orca. I had only seen orca once in my life. It was a quiet day aboard *Balaena*. The sightings had been slow and the weather mild. In my time off watch, I had been on deck for most of the day, transported away from the water and into the dusty badlands of a Cormac McCarthy novel from the onboard library. These books were an eclectic array: a mixture of cetacean field guides, books of knots, and random books that had been left by the various past crews. They were stored on the top of a wet-locker used for hanging oil skins when coming out of the rain. Their pages were wrinkled with the salt air, and flecked with a musky mildew.

Once I got bored of reading, I headed to the foredeck and messed around, trying to get in and out of my yoga poses on the moving boat. I found wheel, and even headstand, but crow pose proved too much of a challenge. Still with energy to burn, I started to climb up the mast and down again until I felt a pleasant burning in my arms. Once back in the cockpit, I let my eyes wander the water in our wake, where the hydrophone streamed out behind. I was rewarded with the little rolling fin of the harbour porpoise, as two of them crossed our stern. My crewmate was delighted. Although he had spent months at sea, all over the world, through his own PhD research, he had never seen the harbour porpoise that I was so familiar with from my home waters. Moments later, there was a palpable change that rippled through the air and the water alike, as a tall dark fin sliced the surface like a knife blade.

Orca. He was identifiable as male from his size, from this

prominent dorsal that towered from the sea like an inky obelisk. As we watched, he was joined by three smaller females. I immediately scrambled back up the mast for a better view of these predators. Whether they were hunting the porpoise or not, the porpoise had vanished. We hadn't seen the orca feeding, nor had a red cloud of blood spread through the water. The large male took an interest in the boat. As I climbed down, he swam alongside the hull. I had never felt such a strong presence before. As I looked down into the water, I was looking at a powerful hunter, and he looked back. It didn't feel like a challenge. Just as a lion would not challenge a lamb, the orca was not challenging me. Although we sailed on, and the orca continued with their own journey, something of that presence remained. The feeling of that day came back to me, though it seemed as though it belonged to a different world, a world in which I had been able to play around with yoga without worrying about a fall, to scurry up and down a mast, to be out at sea and lock eyes with a whale. Now, my only way to find connection with them was through the strokes of my brush.

Christmas came and went, and in the New Year things got worse. I couldn't live by myself any more; tasks like walking and feeding the dog were too hard, and I was missing so many classes that my proximity to university was irrelevant. I clung fiercely to my studies as if they were synonymous with getting better. I was scared that if I had to postpone for a year, I wouldn't have the energy to move forward. I was waiting for my MRI, for which I would probably have to travel back to my university doctor, but in the meantime I needed to be at home.

My painkillers hardly took the edge off any more, my spine had started to swell, so all my medication had to be changed. Anti-seizure medication for nerve pain, codeine as a painkiller, and diazepam to help with the overwhelming anxiety I was experiencing. I was advised that I would probably have to take codeine for a prolonged period, and it was likely that I would become addicted to it, along with the diazepam, and that the doctor would support me to stop taking it when the time was right. I had to go into the hospital regularly for invasive check-ups to see if I had developed central compression in my spine, which would require emergency surgery, and was occasionally given morphine. I had everything pinned on the fact that I would soon get notified about my MRI appointment. In January, over six weeks had passed since I had been referred for the MRI and yet I hadn't received an appointment, so I called the doctor to ask what was going on. The receptionist told me that there had been a mistake, and that the practice had lost my paper-work, so the referral would have to be resubmitted, and that I would have to wait another six weeks.

'Hello, are you there?'

I couldn't breathe, let alone speak, anger taking me to my knees. My mother was looking at me in confusion, as, silently, I handed the phone to her so she could hear the same news. I limped my way outside, pushing open the oak door, faded buoy swinging overhead, and looked down the garden path to the great ash tree, leafless branches swaying against the sky in a light wind. Dusk was fast approaching, the sky already tinged. I felt defeated. It was some comfort to be back in this childhood landscape, but I was no longer that defiant, determined girl who had played on this path, pouring out sugar for ants. My limbs no longer had the strength to climb and swing from that tree, to look at the sea over the roof. My old surfboard leaned against the end of my bed, unused.

My eyes stung, but, once again, I could not let the tears fall. My mother came out into the garden, and I could see she was angry too. She put a steadying hand on my shoulder, her eyes a light grey. She told me that the receptionist had said there was nothing to do but wait, so she had rung a doctor friend who told her she should book a private MRI immediately. She had rung around, and managed to get me an appointment the following week. I had been independent-spirited since very early in life, so it was strange to go back, suddenly and completely, to being taken care of like a child. But I was also grateful, as I knew that it wouldn't be possible to go through this without my family to support me both financially and emotionally.

A few days later, early one morning before the sun had broken the horizon, a car pulled up on the track outside our cottage. I limped down the path, opened the door and got in. It was my friend Joey, and she drove us to a small cove. We sipped hot, bitter, perfect coffee in her car as the dawn brightened, illuminating the stones that had been strewn over the slipway. The sea was not calm, the swell rushing, spray flying, swilling about the bay. Still, we got out of the car, shrugged off coats, jumpers and boots. Joey held my arm, my feet unsteady as we walked towards the water. Rain had started to fall, and the winter cold bit at our skin, already dimpled, turning pink and blue. My legs were shaky. We did not step into the water as such; the sea rushed up to greet us, sweeping us from our feet in a giant, freezing wave, only to deposit us down again, laughing, dripping, shivering and salty. It was probably stupid, the day too rough to swim safely, especially with a damaged spine, but Joey held on to me to steady me. We limped back up the slipway arm in arm, water streaming from our noses. I *felt.* I felt alive again, momentarily cleansed, my body burning from the cold salt water rather than from my damaged nerves. I felt like me. Like the sea.

'Press this button if you're panicking,' the medical staff told me as I slid into the MRI machine. Panicking? My body had been in panic mode for weeks, my brain on a rollercoaster of drugs and emotion, all the while trying to study for modules where the lecturers had yet to see my face. Although I had grown to hate tight spaces, the thing that was worrying me the most about the MRI was a creeping fear that it would show nothing. What if my physiotherapist, my doctors, were wrong in their suspicion that I had a disc problem, and that somehow, my brain had made this all up in some kind of twisted projection because I could not in fact deal with life? I steadied my breath. I brought the feeling of that swim back to my mind. The glorious engulfing of the cold water. How it had lit up my body, made me feel. How it had washed me clean of worry, even for just a morning. I thought of a sea breeze. Of ripples on the water. Of holding my breath, clutching a stone as light streaked and danced through the surface.

My worries were unfounded. The MRI showed that the last disc in my spine, L5 S1, was sagging over the edges of my vertebrae, dehydrated and compressed. This wasn't actually causing the problem, though. The critical issue was that I had developed spinal stenosis, a narrowing of nerve canals in my back caused by calcification as my damaged spine had tried to heal itself. Despite the diagnosis, I was relieved. Finally, someone else could see what I was experiencing: the image of my spine translated my pain into visible form. The consultant told me that I would need surgery. He was confident that I should get 80 per cent or more of the feeling back in my leg, that the pain would subside. My disc would be left alone, but the calcification would be scraped away from my spine, freeing the space for my hopefully not too permanently damaged nerves to recover. Apart from the spinal damage, I was young, fit and determined. I had every chance of getting healthy again. There were risks, of course, as with any surgery, but what choice did I have? The damage was acute, and worsening with time, so my spinal surgery was scheduled for the following week.

The day before my surgery, my mother took me to the water. The February skies were soft grey, muted lavender. Stubby gorse, too early, too cold for bright coastal flowers to paint the landscape. With my shoes off, I squished my toes into the hard, dark earth, cold and moist. I could feel it, the soft soil between my right toes, a quiet grounding thrum of music running through the land. With my left toes, I felt nothing. I looked for fire crow, a sound, or flash of colour, but there was no cawing call of corvid to ring through the air. I stared out to the horizon, the water rippled with a light wind. The horizon seemed impossibly far away, the thin line where sea became sky. I couldn't make it to the sea, the path too steep now, my legs already straining. I felt like I aged with every step, as I limped, way-worn, towards a small waterfall. I knew if that water touched me, it would wash a part of me into the sea I couldn't reach myself.

The air was cold. The water colder. Ice met with the fire of my ragged nerves, and in that instant, I could breathe again, fully, deeply. My head was sharp, clear. The horizon seemed closer. With the water streaming around me, I finally found the tears I had not had the strength to cry in months. They brimmed out of my eyes, mixing with the falling water like a promise. I would get back to the sea. I would sail. I would share breath with whales, feel the tingle of starlight on my skin, the wind behind me and the world ahead.

I woke up, confused, hot, nauseous. This seemed strange as usually I can ease my seasickness by lying down, letting my brain gently adjust. Not this time. I was going to vomit, and, desperate not to do so all over myself, I tried to get out of my bunk. Except . . . no. It wasn't a bunk, and my body would not move. Why couldn't I move? Unfamiliar room. White. A tube in my hand. Another person. A steadying hand on my shoulder.

'Hey, you're awake!'

A nurse. Not a boat, but a hospital.

'How do you feel?'

'Sick.'

She administered anti-nausea medication straight into my cannula. I felt it wash through my body, settling and soothing. Sips of water as I slowly began to fully comprehend where I was, what was happening.

'Did it work?'

'The surgeon will see you shortly, but, yes, everything looks promising.'

I settled. Into morphine, into my bed, finding that I could move more and more as the day went on. My mother came and sat next to me. I felt calm, a chemically induced happy. My surgery had been a success, my nerves freed. I could not believe how I felt as my morphine subsided. I was obviously sore from surgery, but the burning pain that had been my constant, unrelenting companion for almost a year? It was gone. It felt almost maddeningly good, and my relief was deep. I hadn't even realised until now how much it had been in everything. The edge to my voice in every conversation, the dullness in my eyes. In my broken sleep, in the taste of every meal I had tried to eat. I hoped desperately that what had been the hardest, most challenging year was finally behind me.

I have since had a lot of time to reflect on healing. I now know that recovery is rarely like any of the triumphant comeback stories that are portrayed in the media. The reality is a visceral, gritty struggle of progress, uncomfortable lessons and setbacks, and it is unclear when the process ends. For me, there was the physical side of healing. The side that needed surgery. A side that, five years later, means I still need injections in my spine, and I will always need physiotherapy. But there is also the mental side, which is less often spoken about, as well as the part where physical and mental combine. For me, I think, healing is like a garden. You have to choose the right location for the right plants, the place where they can grow. You must provide sunlight, water and nutrients. You must prune with a gentle hand, and rip out any weeds that will inhibit and stifle. It is not a linear process, it comes in cycles, in seasons, with dormant periods. And you need to give it time, which was the hardest thing for me to understand. The pace of my life had always been extreme. I had reached out for anything I ever wanted, worked for it until it was mine. I did not understand how to be slow, how to be still.

As soon as I was discharged, I went back to the cottage with my mother to continue my recovery. I couldn't wait to be there, out of the harsh light of hospital. I needed the familiarity, the walls that were not quite straight, the cold stone floor, and the warmth and light of the fire. I needed the landscape that had made me, and to look out at the sea, vast and complicated, beautiful precisely because I would never truly understand it. But even as my mind wandered out to sea, I was having to learn my way on land again, constantly surprised at the challenges of living in our old cottage. When you are healthy, you do so many things without realising. I could no longer carry in logs from the woodshed, or stoke the fire. I was constantly twitchy that a cat would leap from a pile of books on to me, or trip me as it circled around my feet, and I would pull my stitches. In one of those early weeks, it snowed, and our whole peninsula lost its freshwater supply. Usually I would help bring in buckets of snow to melt, to fill toilet cisterns and to boil, but now I couldn't bend to lift them.

I started to develop extreme headaches. They were debilitating, almost blinding, always worse when I was upright, easing slightly when I lay down. I was worried enough to call my surgeon, who said that there was a chance that a fragment of bone could have caused a leak in my cerebral spinal fluid, and that if I did not see any improvement in a day or so, I should go to the hospital for a brain scan. I hung up feeling as though I had been plunged, casually, into terrifying limbo. I didn't know what to do, so I continued tapping away at the data analysis I was doing for a university report that was due, trying not to think about whether my brain was actually leaking.

The nights were hard. General anaesthetic disrupts the natural rhythms of the body, and now a slip into sleep was impossible. I knew my body needed sleep to heal, but something in me wouldn't let it come. I would lie awake for those endless night hours. The lighthouse beam still swept the land, reaching out over the sea, but instead of taking my curiosity with it into the night, it seemed to taunt me, a lone soul awake while the world slept. The darkness took on a new quality. Nights had always seemed rich to me, velvet and full of possibility. My nights awake used to seem like secret stolen hours, almost as if I was being granted the opportunity for more life. This new dark was an ominous creature, which began to wrap itself around me, tendrils creeping until they poured down my throat and filled my head with terrible thoughts. The early promise that a doctor would be able to help me reduce and stop my medication never materialised, so I decided to do it myself. I popped them all, one by one, white tablet through silver foil, into the bin. I didn't know how dangerous it was to quit them all at once, rather than slowly reduce the dose, but, given the state of my mood, I wonder whether I would have done it anyway.

The nausea came first, lurching and terrible, worse than any seasickness. As I lay, dreamless in the dark, I started to sweat, started to shiver. My hands had been shaking for weeks, but now my whole body quivered. I started to feel a hot scratching in my veins, as if malevolent creatures were trying to claw their way out of me, as my body called out for the opioids and my brain refused to give in. My heart started to race, my breath shallow, as suddenly my mind filled with the overwhelming sensation that this was all entirely my fault, entirely deserved. That I had been selfish in life, that I could never just allow myself to be quietly contented with what I had, that I had always needed more. Why could I not have been quietly happy to live a small life by the sea? That in my urge to always push harder, to know more, to see more, to climb higher, I had, like Icarus, flown too close to the sun. Now the wax was melting, dripping down my shoulder blades on to my damaged spine, and I was experiencing the fall I so rightly deserved. On those nights where the darkness was abyssal, as I lay both hot and shivering, this thought felt like such a certainty I could not see past it. It was lonely and terrifying, and I was wholly convinced that I deserved to be lonely and terrified.

I don't know how many nights passed until the chemicals in my body balanced themselves back towards some kind of harmony, and my thoughts once again became my own. Time has since performed a gracious act on my memory, and the painful edges have blurred. With painkillers set aside, I started to focus on my own breath to manage how I was feeling. As I breathed in the air, I would imagine it filling my lungs, feeling my chest rise, my stomach. I would imagine the oxygen travelling through my body, nourishing each cell before I exhaled everything no longer needed. I breathed in life, I breathed in energy, the world around me. I breathed out all of my hurt, confusion and anger. Part of my mind was sceptical. How could the simple act of breathing replace the medicine that I had been relying on? But as each day passed, I found the medicine in my breath itself. I counted my inhales, my exhales. Always counting, a soft numerical internal rhythm. I could fit three breath cycles between the lighthouse flashes at night. Inhale, exhale. Inhale, exhale. Inhale, exhale, flash. Over and over, as the stars rose and dipped, and somewhere, in the middle of it all, I would find sleep.

My sleeping mind felt so free, occupying a dreamscape where my body was unconstrained and possibility was limited only by the scope of my imagination. The dream I had the first night without painkillers stayed with me, all through the next day, and over the years since. In it, my body was changed, not just because I felt healthy – my arms had become strong, gliding wings. I could feel the tip of each feather as it was lifted by an ocean breeze, and I would shift the feathers in response to keep my flight smooth, gliding. My flight muscles were strong and powerful, but I wouldn't use them, not unless I had to. I wasn't weightless, but I knew my weight, carried my weight so well that I could soar. I could shift, be light, be quick, to keep myself in the air, keep myself on my journey. I had a long way to go, but there was no obstacle which I would not be able to glide over.

The world looked different from on high, the water a glistening, undulating tapestry below me, light glancing, ripples dancing. The air was a world of vertices, each one attainable through the updraughts, currents in the air that pressed against the swoop of my wings and raised me higher. Something in me, something innate, knew the route I had to take. I was a bird, an albatross.

Suddenly, white wings became white sails, flogging in the shifting breeze as the vista reformed around me. I now stood on my own two feet again, the weight of gravity pressing me on to the deck of a boat. I looked around for someone to pull the sheets and set the sails, but I was alone, with the certainty that I would have to set to the task myself. I heaved, I winched, until the boat started to fly, as sure as the bird. Again, I looked around, calling for the captain, but to no reply save the rush of the wind. It looked like I would have to take control. I placed my hand on the tiller and started to bear away from the wind, easing sheets as I went, until I could feel the breeze sitting nicely on my right shoulder and the vessel settled on to the reach. I was at the helm, I was steering my own course again, purposeful and steady on my voyage.

I awoke that morning to the sunlight illuminating my bed, my cat stretching beside me before sliding to the floor with liquid grace to head off in search of breakfast. After my first natural night's sleep in months, I felt awash with calm, but it was the strangest of dreams, and my body still tingled with it. White wings, white sails. Wings to sails. My hand on the tiller. Setting a course, steering a course. Captain of my own ship. It was only weeks since my spinal operation and I was hardly able to walk more than a few steps, and yet I suddenly had a deep certainty that I would do this. Not just that I would sail again, but I would own and sail my own vessel, to voyage when and where I pleased. It wasn't something I had ever previously aspired to, always happy just to be on the water and to crew for others, but now the idea seemed brilliantly important to me. The day I had sat on St David's Head, pondering the environment that had shaped me, thinking how I myself would go out into the world, lit by the pyrotechnic spark of fire crow, I had decided that my life not only needed to revolve around the water, but that I needed adventure too. Once I started sailing, I had needed to satisfy my endlessly questioning mind, to know more about the whales whose space I now shared. I knew I wanted to combine sailing with marine ecology, exploration with research. I had hoped that, on graduating, I would be able to join one of the few sailing cetacean research vessels as a research assistant, or sailing crew. But I had woken up one morning, gone into the sea with my surfboard, and when I came out again, I found that my life had changed, and with it all my plans and hopes for the future. Although I hadn't known it in the exact moment, the last year had brought about a profound shift, and I needed to set my life on a different course. I wasn't yet sure exactly where that would take me, but I did know that I had just been through the hardest year of my life. So why not let this also be a hard thing, sailing my own boat, but the hard thing I had chosen for myself? I felt that the current that ran through my life was gaining speed now that it had a true direction again, and everything in me was swept along with it. Swept along on the wings of the albatross that had flown through my dream.

4

Wandering Albatross

While I was bewitched by dreams of albatross, of captaining my own boat, they were still only dreams, with some very real obstacles to overcome if they were to be realised. I was still spending most of my time in bed, curled up in a duvet with the collie stretched beside me. Her presence was a comfort in the discomfort. I remembered a time on the deck of *Balaena*, in the freezing fog, thinking how nice it would be to be in a big warm bed. Now I would have done so much to be back there. But I could feel myself being healed, letting the nourishment of home surround me. There was still a lighthouse, there were still stars. There was a warm fire, my cats, my books, my family and friends caring for me, a community working together to help me get better.

As I grew stronger, finding confidence on my feet, my ability to exercise and spend time outdoors increased, and drifting off to sleep to spend time in my dreams slowly became a little easier. The muscles in my back were knitting themselves together again, and my nerves were starting to recover from the compression. My left leg was incredibly shaky, the loss of muscle meant that my knee fell inwards, particularly when I was trying to walk uphill, and it was proving difficult to rebuild. A lot of the work was mental, trying to reconnect my brain to my body after the trauma. My mother would drive me down to the beach and walk with me, slowing her pace to match mine; every week I tried to go further, the ultimate goal to reach the south end, where there was no sand left, only water, and walk back again. I wanted to swim again too. But I was no longer the child who could cling to her mother's shoulders, to let her carry my weight as we swam together, and I was no longer the adult who had taken for granted the strength of my body. I went to the pool before I braved the sea, driven by friends and by my father. Although it felt like a pale imitation of the water I craved, it was safer without being tossed by waves. The strangest thing was realising that while my back was hurting so much before surgery, I had grown used to swimming with just my arms to save the pain. Now, I had to remember to add the legs back in. Before long, I felt strong enough to dip in the sea, hanging off a harbour wall so that the salt would wrap its way around my body, a panacea.

Although I had not set foot on campus for a semester, I was well enough to go back to take my summer exams. The experience was surreal. In the exam hall, I felt entirely other. When I was at the beach, in the water or haunting the clifftops, the sea was never still, the sands on the shore were ever-changing, moved by the tide, and so I felt an acceptance of the changes in me. The rigidity of the chair I had to sit in for the exams reminded me of how painful it was to force my body to do something it wasn't ready for.

Once exams were over and the long summer stretched ahead, I began my boat hunt. I was working on healing, and taking any job I could, painting compulsively, endless variants on the albatross of my dream, teaching yoga by verbal cues alone, scraping together what money I could. I knew that whatever boat I could afford to buy would need some work, but I also knew I had gathered knowledge over the years that could be tapped into, whether being able to fix things myself or finding the right person to help. The boat I wanted would have a fibre-glass hull, cheaper and easier to maintain than steel or wood. A boat built for blue water, a boat that would keep me safe. Most importantly, a boat that I would be able to handle by myself. I scoured the internet, bookmarking boat after boat. It seemed to be a good time to buy: there were a lot for sale and prices were competitive.

Over the ocean, down the globe, across the equator, another world was being built. The world of the albatross.

> The world is small. The world is warm. The world is quiet.
> The world provides food, provides safety, provides shelter.
> The world is an egg.
> The egg is in a nest.
> The nest is on an island.
> The island is at the southern reaches of the world.

The egg was laid in December, the austral summer solstice approaching, though summer is not warm there, the highest temperature reaching 9°C. Bird Island, just off the tip of South Georgia, in the South Atlantic. At 54° 00'S, 38° 03'W, it experiences some of the most extreme weather on our planet, including the ferocious winds that whip across the ocean. The island is volcanic in formation, with the biome of the subantarctic tundra. There are both shrubs and herbs here, ferns and clubmoss, with swathes of grassland on the ridges, black rock down to the water, fur seals lining the shore. Bird Island, named for the vast numbers of birds that inhabit this far-flung piece of land, for at least part of the year. Here you will find macaroni penguins, in throngs of up to 50,000; southern giant petrels, with their dark plumage and thick beaks; both black-browed and grey-headed albatross; and finally, on the grassy windblown ridge known to researchers as Wanderer Ridge, the wandering albatross, *Diomedea exulans*. They come ashore here every two years to breed and to lay their eggs.

Our wanderer is a creature of style: long, lightly blushed bill, curved to a hook at the tip, sharp-eyed, dark-winged, speckle-feathered back, and bright-white head. She sits on a hillock of a nest, tufts of grass and clods of earth piled into a pedestal. The air is thick with wind and wing, a soaring symphony punctuated by the clumsy crash as these ocean birds fall out of the air to take their place on land. The males arrived first, towards the end of November, the females shortly after, at the beginning of December, our pair reunited after two years apart on the open water. There is a hum of feather in flight as the birds swoop to and fro, but, for now, our wanderer is still with her purpose. Beneath her, in the cradle of her nest, held close against her, is her egg. Her feathers, so sharp and sleek, insulate her body so well, keeping her heat in. Therefore, in order to incubate her egg, during the breeding season, a patch of these feathers on her underside fall out, leaving a brood patch. It is here, snuggled close against her skin, sharing the warmth of her body, that we find her egg. This egg is the only one that she and her mate will produce and incubate in a two-year cycle. Every so often she shifts, careful not to crush the precious egg with her body as she gently turns it with her bill, ensuring that the chick in the egg does not get stuck to the membrane inside the shell.

Aside from her warmth and this turning ministration, the egg has everything it needs, contained within the boundary of its shell. Once the egg was laid, the cells inside began to form tissue, a complex, calculated dance. They repositioned, divided, divided again. They were assigned specific roles, each one with a purpose. A tiny heart formed and began to beat. The energy to do this, the energy powering this growing life to form and thrive, came from the yolk. A network of blood vessels is formed, branched like tiny trees reaching out for air. The surface of the shell is covered with tiny pores, far too small for us to be able to see with our eyes alone, but still they are there. It is through this network of blood vessels and pores that the egg can breathe, air moving in and out of the shell. As the growing embryo metabolises fat, water is produced and passed out of the shell, again via these pores. As the embryo gets larger, a beak is formed, limbs, feet, the yolk becoming absorbed into the body via the abdomen. As the chick grows larger still, it begins to orientate itself, the head towards the wider end of the shell. Downy feathers form.

What awareness the chick has inside the egg is unknown to us. Does it feel the care, the tender attention of its parents? Perhaps it registers the momentary cold when its parents swap positions, the mother flying off, sometimes as far as Brazil, in search of sustenance to replenish her body; the father settles the egg into his brood patch. The egg, the chick inside, needs two parents, both constantly switching roles as they support each other and the chick through this prolonged eleven-week incubation period. One parent alone would not have the resources to keep itself healthy without being able to leave the nest to forage, and, without incubation, the chick inside the egg would die from exposure. An air cell forms in the shell, near the head of the chick. The beak orientates towards this cell. And then, seventy-eight days after the egg was laid, the beak of the chick breaks through into the air cell, an internal pip as it begins to breathe with its own lungs, followed by an external pip as the shell is cracked and broken from the inside. The egg has hatched, and the world of the chick has grown larger.

Now, the world is a nest.

Our chick meets her mother for the first time. She is more than the warmth that helped the chick to grow through the shell. She is a sleek giant of the ocean air, a bird of many – yet never quite enough – superlatives, with a grace about her even as she sits still on the nest. The downy chick snuggles against her, kept warm against her underbelly as the frigid March winds of the austral summer blow, winter drawing ever closer. All albatross species need the wind, their mastery of its force a part of what they are, but the freshly hatched chick has yet to understand that she will be able to soar thousands of kilometres in a single day with barely a flap of a wing, that her energy expenditure in flight will barely be higher than her mother's is now, as she sits on the nest. For now, she needs to stay warm, and so she shelters, tucked up close.

Two days later, the wind brings her father to her. There is a winged dance above her, as birds come and go from the island. One of them begins to drop from the air, paddle feet stretched out ahead of him to anticipate the landing as he meets the ridge and approaches her mother for a greeting. He is returning from a foraging trip to relieve his mate. She takes to the skies herself, and, with no pomp or preparation indicating a great voyage, she heads out on a search for food that will encompass 3,500 kilometres – such is the sublime scale at which the wandering albatross inhabits the world. For the next month, the parents will brood our chick, continuing to switch places so that one bird will always be with her, the other foraging, raising their chick with their synchronicity. When the voyaging parent returns, they will offer their beak to her, so that the chick can feed on regurgitated oil rendered in the upper stomach of the adult from fish and squid, this substance replacing the yolk which sustained her inside the egg.

I myself was once again preparing to leave my maternal nest. Sitting in the garden under the great ash tree, I was constantly researching boats, calling owners and brokers. When I had a list together, I drove a tiny van down to Cornwall to view some of the boats. Being able to drive again, and to sleep in the back of the van at night, felt like the first stretch of a wing, up into the air, feeling its currents, its lift. It was good to be inhabiting a different space, but I was restless to get back on the sea. So far, my search had been fruitless. The first boat I saw needed way too much work, which only came as a sad surprise to the seller. The others I had seen had been OK, but none of them had the right feeling about them.

I only had two viewings left. I woke up early, a bright July day. The sun was already hot as I climbed from the back of the van and into the front seat. I felt nerves building as I drove over the Tamar bridge, crossing the river where Cornwall meets Devon. These few moments driving over the water transported me back years, to the first time I had sailed under the bridge. There are three ferries that run between Devonport in Plymouth and Torpoint in Cornwall, linking the coastal communities. The ferries are pulled over the river on chains that run under the water, connecting to the shore. You can hear the clink of the metal as you approach by boat, and you have to be very careful not to get too close and snag your keel on the chain. I remember tacking upwind, trying to pick the right moment between the various stages of the ferries loading, unloading and crossing the river. Today, over the sides of the bridge, I could see boats lining both sides of the river before the verdant banks, moored at both ends to stop them from swinging in the tide. I wanted to be down there on the water with them.

In South Georgia, beneath her fluffy natal down, the ocean wanderer is beginning to grow. Her parents breed biennially. In the 'year off', when they do not produce an egg and raise a chick, they devote their energy to moulting a proportion of their primary flight feathers, the remiges. The albatross may be the master of the wind, but in order to fulfil that fabled flight, these feathers are essential, and it is important that they are well maintained. In order to be able to fledge, our young chick must now grow twice the number of flight feathers that are moulted every other year by her parents. Just as the cell division and assignment inside the egg was carefully ordered, so is the growth and development of the flight feathers. These are split into primaries and secondaries. The primaries attach to the manus, whereas the secondaries attach to the ulna. Wandering albatross grow ten primary feathers on each wing, with primary 1 being closest to the body of the bird and primary 10 taking the outer position on the manus, at the tip of the wing.

Primary 1 is the first to begin to develop, appearing when the chick is around 127 days old. Her parents both leave the nest simultaneously now, returning to feed her but leaving her alone on the nest, surrounded by other members of the breeding colony. Fuelled by the regurgitated meals she receives, delivered by alternate parents every two to four days, the growth of the primaries is around 1.4 mm a day. Primary growth will not hit its peak, a rate of approximately 3.9 mm a day, until she is between 145 and 250 days old. As fast as this is compared to the primary maintenance of the adults, it is still slower than the primary growth rates of similar-sized birds, such as the Siberian crane, or wattled crane. There are two theories about why the growth rate is slower in wandering albatross. The first is that if the chick fledges successfully and enters the juvenile stage, it will spend two whole years at sea without replacing the primaries. The primaries are essential for flight, and in this two-year period the bird will travel hundreds of thousands of kilometres, to subtropical waters, where the juveniles learn about life on the wing, life on the water. As stunning as their flight may seem, the voyages of the wandering albatross are not recreational. Their ability to forage is dependent on their ability to fly these distances, which, although they seem extreme to us, are both name and nature to the wandering albatross. The second hypothesis suggests the slower growth rate is due to the fact that the chick undergoes fasting periods while waiting for its parents to return with food. Although they both continue to return to Bird Island to regurgitate nutrients to the chick, the chick may still experience fasts of up to ten days before receiving a stomachful of fish and squid oil.

As austral winter hits the island, our chick not only has to allocate energy to the growth of her feathers, but also to make sure that she keeps warm in order to survive. She is still entirely reliant on her parents for food, while the weather around her grows increasingly inclement. The katabatic winds on which she will one day ride as she circumnavigates the Southern Ocean are the strongest on earth, and they carry with them driven snow which can pile and flurry. She is mobile, no longer confined to her nest, able to avoid these drifts if she must. The temperature on Bird Island will remain below –5°C. Although she is growing her flight feathers, she still has down to keep her warm. She will lose it in stages, feather snow amid the freezing flakes falling from the sky, the down around her head being the last to go when summer comes around again and she is close to fledging. Even the surface of her beak is insulated with a layer of keratin, every part of her anatomy built to withstand these harsh conditions. Although her parents are still rearing her, only 21 per cent of their time is now spent ashore. For the remainder, she is left to hunker down on the ridge with the other hopeful fledglings.

Winter passes, and the world begins to warm again. The chicks that survived the elements wander the ridge, wings flapping often to build their flight muscles.

Now the world is an island.

In the final two months before our chick fledges, the visits from her parents grow increasingly infrequent. She is no longer a ball of downy fluff held snug against a brood patch, but a great thing, heavier now than her parents, mottled with moulted down and her new strong flight feathers, not quite possessing the streamlined majesty of a fully fledged bird. Her plumage is that of a juvenile, her feathers far darker than you would find in an adult. The reduction in feeding as her parents return to Bird Island less and less will help her shed some weight in order to take off. Now December is approaching, and with it the solstice. Almost a year since she herself was laid as an egg. The time for fledging is near, but her final primary, primary 10, is not yet grown to completion. The growth rate has dropped now, from 3.9 mm a day at 250 days old to 0.9 mm a day. She feels a restlessness building, watches the shifts in the weather, and yet she cannot go until this final feather is firmly in place. Other chicks have fledged around her, beginning their own journeys, and she is becoming desperate to follow. Body size will vary amongst wandering albatross fledglings, but wing size remains fairly constant, suggesting that this is the area to which the growing bird directs its resources.

The days are now growing longer, lighter. A stiff breeze begins to blow from the west. Finally, her feather is finished and she is ready to fly. This island is all she has known, and she won't see it again for five years. The closest she has come to flight so far has been seconds snatched here and there as she practised, lifting off into the air on her new wings only to be grounded moments later. But there is something in the air now. Something so familiar to her, the hunger for a place she has never been.

A short run, a swift updraught.

Snatched on the wind, rising, soaring. She takes to the air, no longer a chick, but a bird of legend, leaving the island alone as she fledges into the sky.

> The world is vast. The world is wind. The world roars.
> The world provides food but it must be found, safety
> cannot be promised.
> She is the elements. The elements are her.
> The world is an ocean.
> The world is sky.
> The world is wing, and feather, and wave, and storm.
> The world is freedom.

As I pulled my van into Mount Batten, my nerves mounted. The first boat I was viewing that day appeared to have so much potential, but I knew there was one problem before I even saw it. I had scraped together as much money as I could, and had received some inheritance after my grandfather died, but it was still beyond my budget. When I got out of the car to meet the yacht broker on the slipway, I could immediately feel his judgement in the air. I was summer tanned, with my hair in an icy white pixie crop, and I was thinner than usual as a result of my post-surgery recovery, all of which made me appear younger than twenty-three, which was still young to be buying a boat. I'd also spent the last few weeks living and working out of the back of my van, and had swum in the sea every day in lieu of taking a shower. I must have looked very different from what he'd been expecting.

Despite his obvious misgivings, he proceeded to inflate a dinghy to take us out to the boat on a mooring. We were both silent during the short ride, as I poked at the tube, sagging where he hadn't blown it up enough. He thought I wasn't serious, but I thought something in his manner was keen for a sale. We got to the boat, and I swiftly tied the dinghy on to a stanchion post, then I got to work, going through the boat methodically from bow to stern. I opened every locker, took up all the floorboards, inspected every nook. I looked at sea cocks, examining them for functionality and corrosion, these vital points where water is allowed to pass either in or out through the hull – to let in engine coolant or let out water from a sink outlet, for example. I checked the rigging, both running and standing, to the best of my ability. With the standing rigging, I only knew that it should be replaced every ten years, so I looked for obvious corrosion or damage. With the running rigging, the sheets and halyards, it was much easier to look for signs of wear and fraying. I found my way into every nook and cranny, tested bilge pumps, ran the engine, checked the cleanliness of the fuel, and raised my eyebrows at the erratic wiring.

The paint on her deck was flaky and peeling. Her rails were in desperate need of varnish or oil. Some of the running rigging was down to threadbare, the sheath worn away, the core exposed. Many of her blocks were original and would need replacing. Many of her winches were original too, none of them self-tailing, and some seized. Some lights worked, some didn't, and all her bulbs would need replacing with LED to save energy. Her engine had been replaced the year before. Her mainsail was brand new, stored in the saloon, bright white, yet to be bent on.

She was perfect. Her lines were beautiful. She was sleek and well built. She was exactly the boat for me. The broker's manner towards me had shifted during the course of my exploration. He had started to help me, and even pointed out things that were wrong or less than ideal with the boat as I made my notes. As we rode back ashore and parted ways, I promised I would be in touch soon to let him know what I thought. I cancelled my afternoon viewing, and took myself to Plymouth Hoe.

The Hoe is a long stretch of green space, a bank that slopes down to the buildings on the waterfront, backed on the city side by the Royal Citadel, the Marine Biological Association, and rows of grand houses from the era of the Victorian seafront. From here, you can see all the way over Plymouth Sound, peppered with navigational buoys dictating the channels in and out of the harbours. There is an island in the middle, trees on one side, ruins on the other, Drake's Island, named for the sixteenth-century English privateer Sir Francis Drake, who made Plymouth his home port. The military presence in the city, both past and present, is obvious. On the left-hand side of the Sound, Jennycliff Bay stretches away to the old fort at Bovisand. Warships manoeuvre in and out constantly, sharp grey beasts cutting through the water. The entrance to the Sound is marked and protected by a long breakwater, a lighthouse on one end. If you look beyond it, out to sea, you can make out the tall pinnacle of the Eddystone Lighthouse.

The light marks the Eddystone Reef, a known hazard to shipping since at least the early 1600s. There is no recorded figure for the number of boats that have been wrecked there, but work began on the lighthouse after a merchant ship, ill-named *The Constant*, hit the reef and sank beneath the waves on Christmas Eve in 1695. There have been four iterations of the lighthouse over time. The one that stood prior to the current lighthouse was dismantled, brought back to Plymouth, and is erected behind where I was sitting on the Hoe: Smeaton's Tower, a red-and-white-striped Plymouth landmark. The latest lighthouse is a dark, elegant tower, made of interlocking stone blocks, a genius feat of engineering imagined and realised in the nineteenth century by a man named James Douglas. Years later, I would sail past the Eddystone, on a passage between Penryn and Lymington, before heading out across the North Sea and to the Baltic. I was struck by how the Eddystone Reef would have cut a direct line across my course, had I not known it was there. I could easily see how it might have sunk my boat, and firmly understood the need for a light. For now, I sat and watched the tower in the distance, thinking of the Strumble Head light back home, wondering which lighthouses would play a role in my future, would find their way on to my personal chart of life. The Sound was full with sailing boats that day, as if the whole city had taken to the water. Small dinghies were zipping around on the sea breeze that had built with the warm weather. There were larger yachts, and a few stunning wooden classics tacking around, their heavy canvas sails filled with wind.

I sat and watched the world go by, feeling something like calm coming over me. The next morning, I called the broker. I explained the things I had liked about the boat and the things that would need to be fixed or changed. I gave him an offer, just over half of the asking price, explaining why I felt it fair.

In my mind's eye, I could see him nodding on the other end of the phone.

'Leave it with me. I will talk to the owner.'

By the end of the day, the boat was mine. I was the new owner of all 64/64ths of the vessel – the fractions of ownership into which boats are divided. She was a 1964 Nicholson 32. She was small, tatty and unloved, but she was mine. My own piece of the world with which to explore the seas. The emotional value could not be quantified. They say that it is bad luck to change the name of a boat, but I felt in this instance that fate was on my side. I toyed with names. I had thought to call her *Albatross*, after the oceanic legend, or *Phocena phocena*, for the harbour porpoise I had watched as a child, or *Grampus griseus* for the Risso's dolphin that had always intrigued me. All good names, but none of them fit quite right as I whispered them on her deck. No, I would name her *Brave*, for all that she symbolised to and in me.

I stayed in the south-west for a few weeks, working on her myself, getting help from friends and some professional work done. My back was still healing, and I was struggling to accept the fact that not only did I not *have* to do everything myself, sometimes I physically couldn't. I couldn't lift a heavy fuel can, I needed help to bend on the main. And yet I found I was surprised at the knowledge I had picked up over the years, things I didn't think I knew until I needed them. As soon as *Brave* was ready, I was going to sail her from her mooring in Penryn in Cornwall all the way to Milford Haven in Pembrokeshire. It was the voyage I had dreamed of doing, sailing back into my home waters under my own steam. It wasn't the furthest I had ever sailed – 141 nautical miles in total – but it would be the longest passage I had ever skippered. Two of my friends came to do the voyage with me, arriving late the night before the journey home.

*

Our albatross does not travel with friends, she leaves alone. For all the grandeur of the species and its reputation for epic flight, the reality for fledglings involves a large amount of struggle. Although she has spent months growing and exercising her wings, preparing for flight and flying itself are two entirely different things. Her moment of air and grace is short-lived – she finds herself plummeting towards the surface of the water with all the gravity of her freshly fledged weight. Over the next ten to fifteen days, she will spend hours or even whole days on the surface of the sea, periods of swimming and resting interspersed with flight as she moves northwards. Day by day, she builds strength, technique and endurance.

There seem to be a few reasons why subtropical waters are a favourable destination for newly fledged wandering albatross. The weather is warmer, so the juvenile birds have to expend less energy on keeping warm while they are learning to forage. There is also less competition for food. Of all the birds foraging in the Southern Ocean, the wandering albatross is the apex predator, but although they can outcompete all other species of albatross and petrels, they still have to contend with other wandering albatross. The mature birds tend to forage and wander within the sub-antarctic and antarctic regions, making the subtropics a less competitive place for the juveniles to refine their foraging technique. Of course, there will still be competition for food, but it will be among individuals of a similar stage of development and experience.

Of all the stages of life of a wandering albatross, least is known about this juvenile stage. Fledglings show the same behaviour, navigating into subtropical waters by some unknown mechanism that combines the innate and the learned. She left the nest without her parents. Either one or both birds may have returned to Bird Island, perhaps with a final meal for her, only to find the nest empty, their chick fledged. Her fledging concludes the breeding cycle for her parents. The pair would then have parted ways, each headed to wander the ocean alone, embarking on solo voyages with the promise that they will reunite again on Bird Island in two years' time. If they both return to the breeding colony safely, they will greet each other, an egg will be fertilised, hatched, and the cycle begins again. Although our chick fledged in the same period as many other successful chicks, they do not travel as a flock. She flies solo. Each day, she can fly a little further, a little faster. As her prowess in the air develops, her need to spend time on the surface diminishes. In six months' time, she will have the flight efficiency of an adult bird. That is, if she can survive the next six months.

For the wandering albatross, the highest mortality occurs within the first year of the juvenile period, specifically within the first two months that they spend at sea. Some may never learn to fly efficiently enough to be able to forage successfully. This could be because a chick does not get enough food to build its strength before leaving the island. The wandering albatross largely hunts on the wing, so distance travelled maximises the opportunity for the bird to find food and fuel its body. Or some juveniles may fall foul of extreme weather conditions.

She is both hunter and opportunist. She will catch and kill her own prey, and yet is equally happy to scavenge and feed on offal. Wandering albatross hunt or locate their food in three different ways. The first is by sight. Wandering albatross fly hundreds of miles a day, scanning the sea for their prey. Although we may sometimes regard the surface of the open ocean as a sterile plain, akin to a watery desert, there is an abundance of life there if you understand it and know where to look. Fish and squid make up most of their diet. The Patagonian toothfish is a favourite food, along with squid from three families, *Onychoteuthidae*,

Histioteuthidae and *Cranchiidae*. The wandering albatross consumes a similar amount of squid to the sperm whale, that other majestic ocean giant, although their methods for hunting them differ. While the wandering albatross hunts the surface from above, the sperm whale hunts the depths using sound, and tends to feed on the larger adult squid found there. Wandering albatross are not diving birds, and yet stomach contents have revealed deep-sea fish to be a part of their diet too. How they catch them is not entirely clear, but one possibility is that they feed on deep-sea fish that have died and found their way to the surface. I remember seeing a huge deep-sea squid floating dead on the surface in the Azores; it had been brought up and discarded by a sperm whale, and it certainly would have been within the reach of a wandering albatross.

The second way they locate their prey is through smell. Wandering albatross, like all albatross species – as well as the fulmar, the shearwater and the storm petrel – are members of the order *Procellariiformes*. *Procellariiformes* share a highly developed olfactory bulb, a part of the avian brain that allows the creature to control its sense of smell. The neurones in the olfactory bulb receive information via the nasal cavity, and then transfer that information to the rest of the brain. The olfactory bulb in the wandering albatross is amongst the largest found in any bird alive today. It allows them to search the air for the smells that will lead them to food. You can see this olfactory hunting method in their flight patterns. Scents dispersing over the ocean, with the breeze and updraughts in the flow of air over the surface, do not move via a steady diffusion gradient, where the smell would be stronger at the source and decrease further away. Instead, they disperse downwind, in patchy turbid plumes, with varying gradients of concentration, eddies of scent moving through the air.

Wandering albatross are famed for their flight efficiency. Long before the invention of planes that enabled us to travel the skies, sailors watched albatross, marvelling at how the bird could move effortlessly, with barely the flap of a wing, easily overtaking their ships. The winds of the Southern Ocean are as legendary as the bird. Land masses are sparse in this part of the globe. The winds form a polar vortex and, since they can travel uninterrupted, they build without breaking, earning the epithets the Roaring Forties, the Furious Fifties and the Screaming Sixties, at the respective latitudes of 40°, 50° and 60° S. The waves, without shorelines to crash against, also build, and the water becomes mountainous, an ever-shifting liquid plane of energy. The wandering albatross harnesses these extreme forces of nature with a method of flight known as dynamic soaring. The bird orientates itself directly into the wind. If I were to try and do this in my sailing boat, my movement would stall, and my sails would flog wildly. To sail, you need to be around 45° off the wind to keep moving forward. Every time you tack, bringing the bow through the wind, and setting the sails on the opposite side of the boat, to change direction, you lose speed. The albatross, however, pushes upwards, higher and higher, building potential energy as it rides the wind. Then the bird turns, soaring across and down the wind, potential energy made kinetic under the wing. Once this downward swoop is complete, the process begins again: the bird turns into the wind, rises, casts across and plunges down, resulting in a zigzag flight.

As scents disperse in downwind columns, our albatross will have to track upwind, using this zigzag in order to minimise the loss of ground on the downwind swoop. During the upwind portion of the flight, it is likely that she will lose the olfactory cue of the prey, due to the turbid nature of the scent dispersal, as she moves out of the eddy of scent. The crosswind cast is where she is most likely to regain the trail, before once again turning upwind, moving ever closer to her prey. Being able to use scent to target prey allows her to hunt on dark nights devoid of moonlight, in thick fog or stormy weather when vision becomes obscured. Studies show that around 45 per cent of prey is tracked and caught via scent.

If hunting by sight or scent fails our albatross, the other option is patience. There is a technique known as 'sit and wait'. She will sit on the surface of the water, at rest as she waits for prey to surface or drift into her vicinity. The choice of location for 'sit and wait' does not seem to be entirely random; these wait areas are places where the bird knows prey is likely to surface, either through learned behaviour or sensory cues.

It is likely that many of us, myself included, will never get to watch an albatross in flight. The Southern Ocean is one of the wildest places on earth, and these birds spend so little time ashore over their lengthy lifespans. However, over the past four centuries, the wandering albatross has made a profound mark on Western consciousness. In 1616, the first European sailors rounded Cape Horn, the tip of South America, where the Atlantic meets the Pacific. The route was perilous, but the perilous fast became profitable, first as a race of European monarchs to claim the Pacific, and later as a part of the clipper route and the Grain Race; rounding Cape Horn became the fastest way for goods to travel from Australia to Europe – one hundred days was regarded as a fast crossing. Along with goods traded, stories circulated. Sailors told of how the high winds and waves in the Roaring Forties would suddenly drop, doldrums of ethereal calm would set upon them, and great white birds would follow ships, soaring on the wind like ghosts or the souls of drowned sailors. Even in the early seventeenth century, the albatross was bestowed, or laden with, symbolic meaning.

In the eighteenth century, mysterious stories of the spirit bird reached the ear of the English poet Samuel Taylor Coleridge, probably via a man named William Wales, who was his tutor. Wales was an astronomer who had sailed aboard the flagship of Captain James Cook as he voyaged into the Pacific in search of an undiscovered continent, named Terra Australis Incognita. The hypothesis of Terra Australis Incognita was based solely on the idea that there must be a large landmass in the southern hemisphere to balance the known landmasses mapped out in the northern hemisphere. Although this continent only existed in the imagination, the albatross was very real. The bird left a vivid impression upon William Wales, and then upon Coleridge. *The Rime of the Ancient Mariner* was the work of one year. It is the longest poem attributed to Coleridge, and drew much praise and comment when it was first published, alongside the works of Wordsworth, in *Lyrical Ballads* in 1798.

The Rime's central character is an old sailor, of sparkling eye, who stops a guest on his way to a wedding and urgently begins to tell the tale of a sea voyage he took long ago. Immediately both the wedding guest and the reader are drawn into a world of wildness, of violent seas, splitting ice and thick fog.

We find ourselves caught in a beguiling narrative net, swept away from the here and now and on to a ship in the Southern Ocean. The mariner describes the storm that blows the ship off course and into icy waters, and he describes the moment when, out of the mist, a wandering albatross emerges, calm and serene on the wing. The albatross accompanies the ship for days, swooping and soaring at the stern, and the sailors muse whether it follows for sport or food. Regardless, as if with the grace of the bird, a favourable wind blows steady. But then, without rhyme or reason, the mariner takes his crossbow and lets an arrow fly, shooting down the great bird. A giant stolen from the air. With the falling bird, the fortunes of the sailors are swift to change as the natural world sets against them. The wind stills into a deadly doldrum, not a scrap of breeze to fill a sail and carry them on their journey. Bioluminescence dances across the sea like a malevolent spell. The fallen albatross becomes a physical burden, hung around the neck of the man who senselessly brought down the great creature and sent the world cascading into chaos. Sailors fall dead around the mariner, and yet he cannot escape their punishing, judgemental gaze. The light of day becomes the dark of night, filled with evil spirits who slither and crawl, and the mariner is driven to drinking his own blood, a wicked sacrament as he is starved of fresh water on the briny sea. It is only when the mariner begins to pray and repent that he is saved. The physical burden of the albatross falls from his neck and he is sent to wander. The bird is a connection, the bridge between the natural world and the spiritual, existing in both simultaneously. The mariner lives on, haunted by a search for penance and a compulsion to tell his tale.

I have met my own Ancient Mariner. Although he did not come with a warning to impart, nor a confession, he did captivate me with his glittering eye, and he did have a lesson to teach me. I met him on my first long sailing passage, when I crossed the North Sea from England to Norway. The wooden boat on which we sailed was the perfect setting for our meeting, with thick canvas sails, flaxen rope, a heavy tiller, and masts that had been tarred and tallowed. I am not sure exactly how old he was, beyond ancient. His face showed a life that had been lived, deep valleys etched into his paper skin by time. Most of his hair was gone, but a few downy strands blew in the wind. With age, he had succumbed to fragility, his body breaking, but his mind sharp enough to cut glass. He had been a diver, descending through the zones of the water into the deep, in the early days of the offshore industry. Later, he had sailed in the merchant navy.

I was studying with the Open University, taking an introductory science course to prepare me for when I would start my degree proper later that year. I had an assignment that needed to be submitted the day we were to make port, so, around my duties on the boat, I would find a few hours each day to sit in the saloon at the table with my laptop and books. The Mariner sat with me. He asked me what I was doing, and I told him. He replied that he too was studying with the Open University. He wasn't working towards any specific degree, but would pick up modules of geology here, literature there, followed by history or whatever took his fancy. As I stumbled my way through hydrocarbons and protein synthesis, I would often end up shaking my head with frustration. I did want this – I had chosen to do the course – but it was hard not to feel jealous of others who had time to sit and chat, read or sleep between their watches and duties. I was also finding the work hard in itself, not helped by the fact that I was on a moving boat and a reduced sleep schedule. The Mariner looked at me, and spoke with a broad Norfolk accent. His voice was tired, a whisper of wrecked vocal chords that had been decimated by a lifetime of smoking and the cancer that was to follow, and yet it resonated.

'Don't stop learning, girl. Never stop learning. That is what makes you old. That is what will get you in the end.'

The Mariner never wanted to go ashore. Not in Norway or Denmark, nor Germany or Holland. He had seen them all before, he said, but what I think he really meant was that the ports we called at were not the reason he was on the voyage. He was there because the ship and the sea were more home to him than anywhere he had ever been. He knew that he was dying. He had wanted to come home one last time. At night, he would stand on the deck, hunched in an enormous beige parka, leaning against the dinghy, wrapped up even on the warm, short summer nights. The final night of the return voyage, as we pushed west across the North Sea, it was balmy, calm and dark. The wind had died, the sails were stowed and we were using the motor to bring us home. The ink of the sea and the sky merged in a midnight abyss, and it was hard to tell whether we were sailing through water or air. As the boat moved, the wake shone with the most beautiful, luminescent turquoise. I have never seen such a colour before or since, streaks of magic lighting up the night. I am glad that I got to share this wonder with him. Maybe it was the sea, giving him one last farewell. What was my first voyage would be his last. His ashes are in the North Sea now.

Of course, the real cause of the fall of the wandering albatross is not the result of violence carried out by one man, but of our collective actions. Wandering albatross are as much scavengers as they are hunters, and their keen sense of smell leads them to follow fishing vessels, just as they followed the mariner's ship. As longline hooks are set and baited, ready to catch tuna and Patagonian toothfish, a gaggle of birds collects around the boat. There are wanderers, smaller albatross species, petrels, but the wandering albatross find themselves in the most precarious position. The longline hooks are cast, a trail that can stretch out eighty miles into the sea behind the boat, and the wandering albatross fall upon them, to feed on bait or catch, unaware of the danger. In some cases, the birds escape unscathed. In others, the hooks break from the line and are caught in the gullet of the wanderer, or swallowed, to rust in their stomach, or to be regurgitated to a chick in lieu of a bellyful of food to sustain growth. Worse still, the bird can get caught in the line, or snagged on the hook itself. The great ocean wanderer, tangled and desecrated as it drowns, dragged through the sea on a fishing line. A caught albatross has no commercial value; it becomes by-catch, a kill as senseless as that made by the mariner – worse for the fact that it was not even deliberate.

Being caught as by-catch is a threat faced by all wandering albatross. And it is a threat that is growing as both Southern Ocean and subtropical fisheries expand, as we push into this oceanic southern frontier in order to feed an expanding human population. The juvenile wandering albatross feel this adverse effect even more strongly than the mature birds. While heading into warmer waters may have served young albatross in the past, now it seems that these areas have a higher abundance of longline fisheries in operation. While our young albatross is learning to fend for herself, while she is learning to navigate the world alone, her chance of becoming by-catch is higher than that of an adult bird. Coleridge knew nothing of Southern Ocean fisheries, and words like by-catch and anthropogenic climate change did not yet exist, but his poem has echoed over the centuries in a way that is increasingly haunting.

The passage home aboard my new boat started just after breakfast, catching the morning tide out to sea. We headed out through the Carrick Roads, towards Lizard Point. The sea around the Lizard headland forms a tidal race as it stretches out to the south, but I had planned the voyage so we could use that race to our advantage, gaining speed. The weather was fair, the wind light. Checking my watch, I decided to nip a little closer to the land, where the race was running at its strongest to catch the last of it. It was a good call. No sooner were we past the lighthouse that stands on the point than overfalls began to whip up behind us as the tide changed; anyone still in close to the headland would experience a bumpy ride. Land's End, the most westerly point of England, was the second and last tidal gate of the voyage. Once we had rounded the point, we would turn north and sail for an uninterrupted 100 miles. I had only rounded the point once before, and that had been heading in the other direction. This made me nervous. The tide there is incredibly strong, the inshore waters around the point are littered with rocks, and, if you run too far offshore, you risk entering the shipping lane, with tankers bearing down on you at twenty knots. An advantage was that the flow of tide for rounding Land's End is far more favourable when heading east to west rather than vice versa. I held my pilotage plan in hand, and checked off cardinal marks as we passed, reciting their names under my breath, Carn Base, the Runnel Stone – a strange Cornish prayer. As soon as we rounded Longships Lighthouse on Carn Bras, bathed in the evening sun, the hardest part of the voyage was over. The wind had died entirely and we had to motor, but the surface of the water was smooth and gleaming, an unending stretch of quicksilver on which we could glide for ever, or so things felt. Maybe we could forget our destination, all our plans, and just continue on, over the Atlantic on a pillowy sea.

Instead, we stayed our course, heading a few degrees shy of due north. As we switched position on deck through the night, the journey remained smooth. Occasionally we saw the lights of a fishing boat in the distance, but, for the most part, we had the world to ourselves, a faint glimmer of shining phosphorescence in our wake. Day broke, and after we shared a coffee in the cockpit, I wandered up to the bow alone. *Brave* was cutting sleekly through the water, there were teasings of fog around, but nothing that would not burn off with the rising sun. As we entered Pembrokeshire waters, we were greeted by puffins, razorbills and guillemots paddling on the surface, ducking under the water at our approach. The sleek back of a minke whale rolled at the surface before disappearing back into the blue. On our approach, we encountered a basking shark, the first one I had seen since childhood. It had a triangular fin raised above the surface, mouth gaped wide as it filtered the water for food. I pointed it out from the bow, but it was quick to flit out of sight. To finish off the journey, we were escorted by bow-riding common dolphins, revelling in our wake as we entered the Cleddau River to head to the marina and dock. It seemed as if the very shores were welcoming us back, as if in acknowledgement of what that journey meant to me.

For all the triumph of that passage home, it didn't herald a return to the life I'd had before the injury. Despite the fact that things were going well, I often felt like I was waiting for the next disaster. Until my spine was injured, I had no experience of anxiety, but now I carried it with me, and I struggled to articulate what I felt, let alone find a way to deal with it. In my head, I had made sailing synonymous with healing, and although I am sure that in a way it was, it was only a piece of the puzzle rather than the whole picture. In my early dream, when I felt my hand on the tiller, I had felt that this alone would give me direction over my life again. Like the juvenile albatross on her first fledged flight, I had been expecting to spread new wings and ride away on the ocean currents, but there were times when I crashed to the surface and was forced to swim for days at a time rather than fly. I used forward momentum as a coping mechanism, which looked like progress or achievement, a strong bounceback from a bad situation. In reality, movement of any kind was the only thing that was keeping my mind from wandering off a precipice and into chaotic darkness. The sea was never still. It was the place I had to be.

One evening, unable to settle or eat, I slipped my lines, headed through the lock and out for a solo sail. The end of the summer was close, the light drawing low. Instead of feeling sure and steady on the tiller, there was a shake to my hand. I watched the shadowy shapes of the land as I passed buoy after buoy, heading out down the channel to the sea. Where was I going? Why? Who sets sail while dusk is falling, alone, without a destination in mind? No, sailing had not been the cure-all, of course it hadn't. What I needed was time and stability, which seems obvious in hindsight, but was hard to grasp in the moment. Suddenly, I noticed a haunting white shape on the water ahead of me, a physical manifestation of what taunted my brain. My pulse quickened and my heart raced, my hand gripping the tiller until my knuckles whitened. This poltergeist started to splash at the water, churning furiously, cawing as it lurched towards me, and I screamed. The scream turned to a shaky laugh. A gannet. It was only a gannet, probably a fledged juvenile whose sleep I had disturbed with my night-time wandering. I pushed the helm over to turn around. It was clearly time to head back to shore.

If our wandering albatross survives, if she grows strong against the storms, if she learns to find food and avoids drowning on a line, if she reaches five years of age, it will finally be time for her to bring her world back to the island. Wandering albatross breed on a circle of islands scattered in the Southern Ocean. Breeding colonies are also found on the Crozet Islands, the Kerguelen Islands and the Prince Edward Islands in the Indian Ocean, and on Macquarie Island in the Pacific. Despite all this scope and scale within her reach, when our young albatross goes to the shore with thoughts of breeding, she will return to her own natal colony on Bird Island in the Atlantic. By some unknown compass, she wings her way back there, arriving alongside the rest of the breeding colony. Previously mated pairs gently touch bills and delicately preen the feathers of their mate, reunited after their 'off year' from breeding. The surviving juveniles from her cohort will return, as well as the birds born one and two years before her, with the intention of finding mates of their own.

The unmated males gather together, slightly out of the way of the breeding birds, who may be copulating or already incubating eggs. These young males begin a display designed to catch the attention of the young females. They bow their heads and snap their beaks. They shake their heads quickly from side to side. They stretch out the length of their wings, they send their heads jutting forward, long necks rigid, beak outstretched. The male dips his head low towards the earth, and, with a jerk and a low gargle, he quickly flicks his head up, bill stretching up into the sky. While he is conducting this dance, the female circles overhead, flying past as he attempts to secure her attention. It seems to work, because as he points his bill into the air, she swoops low towards him, and reaches out her bill to lightly touch his. She comes into land on the grassy ridge a short distance away from this favoured male, and begins to walk towards him with a slow, swaying gait.

These displays will continue, the young male and the young female bowing to each other, snapping their bills, dipping their heads, bill-fencing. If she is unsure about the male, she will leave and the process will begin again. In this case, though, it seems the pair like each other. They won't breed that year, or the year after or likely the year after that, although they will both continue to return to the breeding colony annually, courting each other and strengthening their pair bond. She intends to mate for life, and will only switch her mate if they consistently fail to produce a viable egg or raise a chick together. On the year when she finally returns to Bird Island ready to copulate and lay her egg, she needs to be sure she has chosen her partner carefully, and that they can support each other through the extended period of incubation and chick-rearing. Once the egg has been laid, she needs to trust that her mate will return so that they can switch places, so that he can take a turn incubating and she can head to sea to nourish herself with food and water. When the chick is hatched and old enough to be left alone, the mated pair need to be in synch to maximise the efficiency of their independent foraging flights. She doesn't want to have to cut a flight short to return to feed the chick only to find that it has just received a meal from the male. Nor does she want to return to find the chick undernourished.

Three years after her first return to Bird Island, our albatross arrives back on Wanderer Ridge at the beginning of December. She is now eight years old, having lived just under a fifth of her natural lifespan. It is time for her to breed. She greets her mate, her now familiar companion on land. He arrived slightly earlier than her, and has secured a grassy pedestal nest. They rub beaks and gently preen each other's breast, a low gurgling sound coming from each bird. The time has come. The birds mate, an egg is fertilised and laid, an egg which she holds close against her brood patch, just as her own mother did, her body heat keeping the egg alive. Inside the shell, the complex dance of cell division begins as, once again, life starts to form. She gently turns her own egg. She sits. She waits, with a stillness she has never before known, as the embryo develops underneath her. A little over a week passes, and she finds that she is beginning to watch the sky, looking for the return of her mate. When he does come back, they carefully swap positions, and she takes to the air. This is the first flight she has ever been on where a strong thread ties her to land, the need to support her mate, to raise her chick. She flies swiftly, purposefully, from South Georgia to Brazil, feeding all the while before banking and heading back to her mate, to her egg.

She sits, she waits, she turns the egg. Days pass, weeks. She sits, she waits, she turns the egg. She is growing hungry, thirsty. She is devoted to the egg, but she keenly scans the skies. Where is her mate? She sits. She waits. She turns the egg. She can feel herself growing weaker, feels that she is losing condition, and still her mate has not returned. A day passes, and another. She is growing frantic. Her basic instinct is to incubate her egg, to raise a chick who will fledge as the next generation of wandering albatross. But there is a more powerful instinct within her. It is the instinct that kept her alive as a chick through the harsh winter conditions. It is the instinct that helped her learn to forage, to perfect her flight so she could travel further, faster. Her instinct to survive.

Some 1,500 miles away, a fishing boat is hauling in its lines. Snagged and tangled, broken wings wrapped in line, her mate is hauled on to the deck. He is drowned. He will become a statistic, if he is reported at all, drowned as by-catch. Accidental. Of no commercial value. What harm can the destruction of one bird do? Back on Bird Island, she stays on her nest as long as she physically can stand. Her chick would hatch in a matter of days, fluffy heads are already bursting out of the nests around her, but she has to go. She cannot wait any longer. Her body, despite the lack of food, has never felt heavier. Her wings know what to do. The wind snatches her into the sky, carrying her away on a mournful journey of lament over the Southern Ocean.

The egg, an entire world encased within a shell, inside a nest, on an island, is left behind, cold and abandoned.

Our albatross will never know what happened to her mate, why he failed to return to her after all those years of careful courtship. It is possible that she will seek a new mate. Researchers on Bird Island say that sometimes they see a bereaved bird find a mate again within a year. For others, it takes several years to go through the grieving period. And there are some bereaved wanderers that will never mate again. Our albatross will not have been the only bird that season to lose her mate and be forced to abandon an egg or a chick. The breeding population of wandering albatross in South Georgia has been monitored for over thirty years. Between 1999 and 2018, the number of breeding pairs recorded on the archipelago has declined from 1,182 pairs to 661. On Bird Island, the decline is 3.01 per cent a year. Although breeding colonies of wandering albatross in the south-west Indian Ocean have seen some recovery since the 1980s due to changes in fishing practices that reduce albatross by-catch, the population on South Georgia shows continued decline. In just nineteen years, 521 breeding pairs were lost. If this rate of decline continues, in little over nineteen years' time, there will be no wandering albatross left on South Georgia at all. The decline in breeding pairs is the consequence of both mature and juvenile mortality. Sometimes one or both members of an established pair are lost. Sometimes, juveniles never live long enough to return to their natal colonies and find a mate.

The wandering albatross has fascinated me since I discovered Coleridge's poem at the age of fifteen. Later, it was part of the opioid dreams of my recovery from surgery, a strange link to the addiction that consumed Coleridge's life. For me, it became a totem, a symbol of hope for those who find within themselves a ranging nature. In many cultures, the bird also has symbolic significance: it is associated with freedom, strength, and wanderlust. For seafarers, the sight of an albatross was thought to bring good luck. In Coleridge's poem, the nature of the luck brought by the bird changes in response to human behaviour, turning from good to bad. For the Ancient Mariner, when the albatross was destroyed by an arrow from his bow, the natural world itself immediately rebelled. The wind stopped blowing. No rain fell to provide drinking water. For us, the harm we bestow on these ocean birds through our consumption happens out of sight, and the repercussions of our abuse of nature are not as immediately obvious. And yet the slow, careless destruction of this noble species is a burden of shame that hangs around humanity's neck. Like the mariner, we still have time to repent, to recognise that these birds share with us an instinctive curiosity about what is over the horizon, a desire to wander and a compulsion to return.

5
Humpback Whale

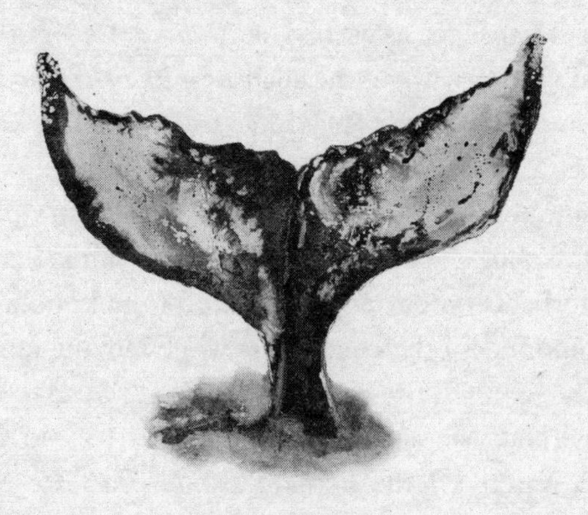

Summer had brought its usual haze and hustle to the shores of Pembrokeshire. Despite having bought *Brave*, after the long sail home from Cornwall my time on the water was fairly restricted. I took her out whenever I could, often in the evenings, sailing down the Cleddau to Dale, where you could drop anchor and swim. I always chose a certain area I knew was free from sea grass to drop my anchor, so that it didn't claw through the marine vegetation. I had two university exams to catch up on. My memory had always been sharp, but since my prolonged drug use, the surgery, the recovery, I was struggling with retaining information, and had to rework all my study methods. While trying to memorise my notes, my mind would suddenly flash back to the more unpleasant points of the previous year, the marine information far too entwined with my health struggles. I would have to stop, sitting in the saloon on *Brave*, my breath short, a clammy sweat breaking out all over me, until I could will myself towards calm. I was also planning my dissertation project, the fieldwork which I needed to complete before the next term. I wanted to take *Brave* out into the waters off the Pembrokeshire coast, to sail transects, as we had on *Balaena*. Except this would be a visual survey, without the acoustic survey of a hydrophone, which seemed a step too far for me at this stage. I knew from working as a tour guide that there were minke whales, fin whales, humpbacks, common dolphin and Risso's dolphin in these waters, but all of these previous encounters had been in noisy motorboats on day trips. I wanted to sail out, away from the land, and record the encounters while moving quietly with the wind.

Even without the acoustic element, this was an ambitious idea. The waters off the Pembrokeshire coast are extremely tidal, and there are many a rock and reef to run you afoul if you don't know how to find your way. There aren't many safe harbours, safe anchorages, if you need to retreat from the weather. However, I had been learning these waters for years, they were my home. They were the waters I loved most, and I wanted to find more empirical data on how and where cetaceans inhabited them. I had a particular fascination with the Risso's dolphin. Although they were a little trickier to find than common dolphin, who often approach to bow ride, I had seen them while I was a guide, and my friends who still worked in that industry told me that the sightings were becoming more regular. I had read studies that had focused on using photo ID for the Risso's, who can be identified individually by their scarring patterns. As they had recorded thirty-two sightings of mothers with small calves present off the Pembrokeshire coast, it seemed that these waters could be important calving or rearing grounds for these dolphins. I wanted to add data that could help support this.

First, though, I had to make good on a promise I had made to my mother last year: to take my Yachtmaster exam. The Yachtmaster is needed to gain a 'certificate of competence' in the UK, although it is recognised internationally. It is split into levels – coastal, offshore and ocean. You can take the exam purely for your own confidence and education, but you also need it if you want to work commercially beyond a certain level.

When my mother asked me to do it, I felt a little hurt. I wondered why she didn't think I could sail, why she needed someone else to sign an expensive piece of paper to confirm what I could tell her. Her retort was so calm and convincing. She explained that she simply didn't know what it was like when I went to sea. Sailing at all, let alone sailing offshore, was so far from her experience, just as the world of art had been far from her own mother's. It wasn't that she didn't trust me, for she did, but, just in case anything ever went wrong, she wanted to know she had done everything she could to see that I was safe.

The female wandering albatross does not fledge with her chick to show it the ocean. She provides provisions, meal after meal brought to the chick, so that it can grow strong enough to fly, so that it can explore the Southern Ocean by itself. I also appreciated that part of the beauty of the sea, the adventure of sailing, lay in the fact that it was dangerous. Dangerous and to be respected. With this in mind, I booked the exam for offshore sailing. I had been building the requisite miles and experience for years, and although I had taken the theory part of the exam, I still hadn't done the practical.

Two days before the preparation for the practical began, while

I was scrambling to renew my sea-survival and first-aid certificates and get a private medical exam, I received a phone call. It was Richard McLanaghan, one of the co-directors of Marine Conservation Research, the non-profit organisation that runs *Song of the Whale. Song of the Whale* is a custom twenty-one-metre sailing boat, built to conduct sailing cetacean surveys anywhere in any ocean. The project has been running since 1987; the current vessel is actually *Song of the Whale II.* The first boat and crew played a crucial part in helping the Azores transform from whaling islands to whale-watching islands, demonstrating that these creatures have greater value alive than dead. Some years ago I had made contact with Richard and I had gone aboard *Song of the Whale* to meet him and talk about the possibility of gaining experience on the boat. Now, it seemed that my name had come up again. *Song of the Whale* had been conducting a cetacean survey spanning the entirety of the Mediterranean for the Agreement on the Conservation of Cetaceans of the Black Sea, Mediterranean Sea and Contiguous Atlantic Area, or ACCOBAMS. The survey was made up of acoustic transects and distance sampling to determine the abundance and distribution of cetaceans in the Mediterranean. The crew from *Song of the Whale* had been joined by scientists and conservation workers from the countries that bordered the survey areas, from both the Mediterranean and North African coasts. And now they were looking for temporary crew for the final stage of the survey.

Richard asked if I would like a position on the boat, and I immediately said yes. There wasn't really anything to consider. Between buying *Brave* and booking the Yachtmaster, my bank account was almost empty. I didn't have enough money to cover my next term at university. My reasons weren't just financial, but the need for money did help me to put imposter syndrome to the side. For years, I had considered *Song of the Whale* my dream job, and I had also doubted I would ever get there. She is a lot bigger than *Balaena*, 68 foot to *Balaena*'s 40 or *Brave*'s 32. She's category 0, designed to be able to sail anywhere, from the balmy Mediterranean, to Greenland, or the Southern Ocean, where the wandering albatross make their flight. She has four cabins, a huge saloon, a fully equipped galley, a workshop in the lazarette, and – the real novelty for me – two fresh-water showers on board. Not only is she designed to go anywhere, but she is made to be at sea researching for extended periods of time, without the need to call ashore. Getting to spend time at sea, sailing commercially, and studying cetaceans meant I would be exploring and giving back to the sea simultaneously. It seemed like the ultimate privilege. Richard seemed happy, but, before he hung up the phone, he had one final question. My Yachtmaster. Did I have it?

'I am starting it in two days' time,' I said. 'I should have it by the thirty-first.'

Should have it by 31 August. From there, I would fly out to Crete to meet the boat on 2 September.

On 27 August, two days after my twenty-fourth birthday, I began my prep. There was only one other person on the course taking the exam with me, a man in his forties, but the crew was fleshed out with two women who were learning to sail. All of the prep, and my exam, took place in Plymouth Sound and the surrounding rivers, so I went from sitting my catch-up university exams straight on to the water. The first three days were a blur of drills, night navigation practices and passage planning. On the fourth day, when the examiner stepped on the boat, my hands were shaking and I had to force myself to take deep breaths. It was time. The minute the exam started, my nerves dissipated somewhat, as I began to see it as a series of challenges to overcome, one at a time. We had to sail without the use of GPS. Although these days it is rare that you will be in a situation without GPS, it is important to be able to navigate without it. Since the weather was light, most of my tasks were manoeuvres under sail, and navigating up the Tamar in the dark on a falling tide. The Tamar River, flowing from its source at Woolley Moor out into Plymouth Sound, is tidal, with large swathes of water drying to mudflats on the ebb. In the early 1800s, river barges were built specifically for moving goods up and down the waterway. They were designed with a shallow draught, so that they could travel as far as possible, even while the water fell. I know of one such barge that is still sailing, the *Lyhner*, named after a tributary of the Tamar. These days, she carries charter guests rather than goods. Unlike the *Lyhner*, the boat I was sailing in had a reasonably deep keel.

I had to be careful not to run aground, checking my tidal calculations, constantly looking at the time to make sure I had enough water. The night was very dark, the hour late, as we had had to wait for the late-summer sun to set. Large stretches upriver are buoyed but not lit, so I couldn't rely on the flashing navigational lights to guide me. I could make out the riverbank, but only just. I could hear curlews, small cries in the dark. I could picture them, waiting as the water fell away to silty mud to begin to feed on the invertebrates they dig from this intertidal zone. I used another sound to find my way, beyond the call of the waders. A depth sounder on the bottom of the boat was sending out a signal, and as it was returned the depth of the water was relayed to me. The navigational charts I had been given showed a five-metre contour line, running up the Devonshire side of the river. The depths on the chart show datum, the lowest astronomical tide. I had calculated how much deeper the water should be above datum, for the specific date and time of our navigation, and used that line to find my way. If the readings from the depth sounder started to become shallower, I was getting too close to the bank; if they became deeper, in the middle of the river, I could adjust my course. Once I found the requested point, I was then asked to go below decks, to repeat the navigation back to the mouth of the Tamar totally blind. Seeing as it was sound that had helped me, rather than sight, it wasn't much more of a challenge, beyond the need to be extremely clear in communicating with my crew, who remained on deck, while I was at the chart table. As we wiggled our way back down the dark river, I was constantly calculating, using our speed and the time to work out how far we had travelled. I had been asked to stop just before we reached the chain ferries that crossed the river. I told my examiner, who was sitting in the saloon across from me, watching my every move, that we were there. He asked if I was sure, and I said I was. He asked if I wanted to check my calculations again. I didn't. It was sound that had given it away, rather than my fluency with numbers. When I heard the familiar clink of the chain, pulling the ferry across the river, the noise carrying easily through the quiet night, I knew exactly where we were. By the time we docked, it was 3 a.m.

The next afternoon, after some time spent manoeuvring in a marina, I went to the office of the sailing school for some final questions and to hear my verdict. I had passed. The examiner asked me how I had found the experience, and I didn't quite know what to say. It had been intense. For the past fortnight I had exhausted myself, constantly running over potential scenarios that would occur in the exam, and how I could solve them. I had worried that I would be exposed for trying to be something I wasn't and sent home. But then, once the exam had started, I had loved every minute of it. Of course it had been challenging, but I had found my mantra once again. It was a hard thing, but I had done hard things before. On hearing my pensive answer, the examiner told me that he had just given me the most rigorous exam he had ever done. He said he had been intrigued by my response to stress, and had thought that he would be able to push me beyond what I believed I could do. I have thought about what happened a lot. He may have taken a risk, but it was exactly what I needed. For a stranger to have seen and understood my capability was a compliment that has given me confidence in challenging situations I've faced since.

I shared a final drink with my crew before heading back to Pembrokeshire to tell my mum the good news, and make sure that *Brave* was tied up as safely as possible, my father checking on her in the weeks I was to be away.

Two nights later, I was walking through the hot night air in Chania, Crete, to join *Song of the Whale*. In a few days, we would set sail for the Libyan part of the survey. While we waited for the rest of the crew to join us, I started getting myself acquainted with the inner workings of the boat. As I familiarised myself with the decks, a seal swam into the harbour, inquisitive eyes watching me as I went about my work. I asked someone on the boat next to us, a wildlife trip boat, what sort of seal it was, and they excitedly said it was a monk seal. It was light in colour, silvery, more like the young Atlantic greys I knew from home. Its face was much shorter, rounder. I didn't know much about this phocid, and, having grown up in Pembrokeshire, where seal sightings are common, I didn't realise how unusual this one was.

Later, I learned that the Mediterranean monk seal is the most vulnerable marine mammal in Europe. Just like the Atlantic greys, the Mediterranean monk seals are vulnerable to by-catch – they get tangled and drowned in nets – but the fishing pressures here in the Mediterranean are far greater, decimating the population. Fishermen have also killed the seals, so as to not have to compete with them for fish. The pressures of tourism in these waters are higher. Although sometimes Pembrokeshire can seem full to bursting in the summer, there are still secluded bays and nature reserves where the mothers can pup in peace. Here, people are encroaching on every aspect of the monk seals' lives, from what they eat, how they hunt, to where they can give birth to their young.

In the afternoons, when the temperature reached sweltering, I would take a break, climb over the harbour wall and go for a swim. It was the first time in my life that I had swum in warm water. As a child, I had imagined how it would feel to slip into water that did not have the bite of the Celtic Sea. It was blissful and relaxing to swim in this balmy warmth, fish flitting over the rocks beneath me, and yet it also made me glad I came from somewhere where the sea held more challenge, where comfort was found by expanding your limits.

When we set sail from Chania, passing out beyond the harbour wall and the stratified cliffs that make up the island, the winds were strong, blowing hot salt air across our faces. The first period of time with a new crew is always a little odd, as everyone takes a while to adjust to a new rhythm. Later in the survey, when everyone was familiar with the timetable of the watches, it would be rare to find everyone on deck all at once, except during dinner time. For now, though, the cockpit was crowded, everyone preferring to be above deck to counter the early nausea.

The days were hot. So hot in the afternoons that I would often have to pour water over myself, which quickly dried to white salt crystals on my skin. My small bunk soon felt like home, and although I was still not sleeping very easily, I would plug in my headphones and listen to stories while I rested. For the first time in months, my mind was no longer trying to work three steps ahead. I was content to just be where I was, present on the boat, all thoughts of the shore behind me. I shared an hour of my night watch with the skipper, and an hour with one of the Libyan crew members. As we sailed towards his home waters, we would talk to each other, him enquiring about my life, me equally curious about his in Libya. I had at that point only lived in the UK, I was a woman in my early twenties who, for the most part, lived alone. I travelled, I studied at university, and here I was, at home on the sea, working in an environment that was familiar to me. During my lifetime, he had experienced the first Libyan civil war, and was currently living through the second. He had never been on a sailing boat before, and asked the most wonderfully inquisitive questions about how sailing worked. He told me how he had loved to dance as a boy, but now he was grown, and there wasn't time for dancing. Somewhere between the Mediterranean and North African coasts, under the night stars, we decided that there was no time like the present, and he moon-walked up and down the cockpit. The crew was made up of people from Wales, England, Ireland, Germany, Spain and Libya. We all had entirely different backgrounds, and yet we were all brought there in one way or another by a love of the sea.

The most distinctive aspect of the research on *Song of the Whale* is the use of sound to explore the sea. Through the hydrophone, I could hear the sound, from deep below, of raindrops falling on the surface; the whistles and clicks of dolphins; the louder, metronomic click of a sperm whale; the sounds that *Song* herself made as she travelled through the sea – the melody of moving water. Through sound, we were able to enter the world of the whale. Many of us hear so many sounds throughout our days that we begin to tune out, to let them become background. We create the pictures of our world largely through sight. It is easy to hear sound without ever really stopping to think about what it fundamentally is.

Sound is an auditory sensation, a mechanical disturbance. Once the sound has been made, it must then travel through a medium. In our terrestrial lives, this medium is generally air. In the sea, it is of course water. The vibration of the sound body occurs first, and the propagation of the sound comes second. If someone calls your name from across the room, the vibration starts with their vocal chords. Their mouth shapes the sound into a word, and the sound is then propagated through the air. From the sound, we are able to process and ascertain information. Your friend called your name to attract your attention, and you can choose whether or not to respond to the stimulus. You also gain a sense of directionality from the sound – you know where your friend is standing, in relation to yourself, and how far away she is. The sound she makes can be measured, in both frequency and amplitude. Frequency is measured in hertz (Hz), and quantifies the number of cycles of the sound wave that pass through the medium in one second – the pitch. Amplitude is measured on the logarithmic decibel scale (dB) and describes the pressure or forcefulness of the sound – how loudly we interpret the sensation. Sound is useful on land, but in the sea it is essential.

In the sea, light fades. Below 200 metres depth, we enter the twilight zone. Light attenuates more in water than sound, so, although the world grows dark, sound can still travel. Sound really is the sense of the seas. The first depth soundings of the deep ocean were taken in 1839 by Sir James Clarke Ross on board one of the two ships he was commanding, bound for the Southern Ocean. He used a lead line, 2,425 fathoms (or almost 4.5 kilometres) in length, in order to reach the bottom. I cannot imagine how the line was stored on board, deployed and handled by the sailors to gain an accurate measure while at sea, but they were using the tools they had. The uses for sound underwater gained a new urgency with the sinking of the *Titanic* in 1912; the disaster prompted a commercial race to create a device that could detect objects under the water. In 1935, the first two patents were filed for what would come to be known as depth sounders, the same as I used to find my way up the Tamar. The idea was that as ships went along, they would send a sound signal out into the water. From the time it took for that signal to return to the receiver, the distance to any objects in the water could be calculated. It was assumed that the sound waves would travel through the water in straight lines, so the signal would always go direct to an object, and bounce straight back.

This is not the case. Sound travels through surface water at an average speed of 1,500 metres a second. But this is only an average, and the speed of sound is affected by many things. Let's first consider the temperature, which decreases with depth. The mixed surface waters are warmer, the deep ocean colder, creating a thermocline. As the water temperature decreases, so does the speed at which sound travels. What's more, when a sound wave enters an area of different temperature, it not only speeds up or slows down, but refraction occurs and the direction of travel is altered. However, hydrostatic pressure and salinity will also change the speed at which sound travels through the sea. The deeper you descend into the ocean, the greater the weight of water above you, and therefore the greater the hydrostatic pressure. This increase in pressure counterbalances the effect from the decreasing temperature, leading to a sound wave constantly shifting in both speed and direction. Although sound waves travel irrespective of the flow of current, any turbidity causing air bubbles in the water will alter the speed of sound, as the sound wave temporarily passes through a medium of air rather than water. Although sounders were designed to establish measurements of depth and detect objects below the surface to minimise the risk of physical collision, there is more to the sea than the seabed and icebergs. Sound signals are returned from the bodies of fish, of whales, of plankton, of squid, and a whole host of marine life.

In 1946, three scientists, C. F. Eyring, R. J. Christensen and R. W. Raitt, reported a discovery after conducting acoustic surveys off the coast of California. They described a layer, around 300 metres in depth, in the water between the surface and the bottom. This phenomenon of the 'false bottom' turned out not to be limited to the Pacific, but was found in all ocean basins. What's more, it was not stationary; it appeared at varying depths depending on the time of day or night. This layer was initially named the ECR layer, after the men who discovered it. In her 1950 book, *The Sea Around Us*, Rachel Carson discusses the three predominant hypotheses about the formation of the layer that were circulating at the time. The first was that the ECR layer was planktonic, the second was that it was made of fish, and the third was that it was composed of squid. When I came to study at university sixty-five years later, I learned of the ECR layer as the Deep Scattering Layer, or DSL, named because of how it scatters sound.

As the nature of the sea is to never provide one simple, straightforward answer, the hypotheses Carson described all proved in part to be true. This 'false bottom' of the ocean is caused by sound bouncing off zooplankton, the swim bladders of fish, squid and other marine creatures. Zooplankton, the plankton made up of animals and protists, reside in the deeper, darker water during the day. In darker water, they are harder to see, and therefore less vulnerable to predation. They have to rise towards the surface between the hours of sunset and sunrise to feed on phytoplankton. Phytoplankton are photosynthetic, and therefore need the light of the sun in the surface waters to generate energy. As the sun rises and visibility in the water increases, the zooplankton once again descend. The zooplankton provide a link in the food chain. As much as they try to minimise their chance of being eaten by larger marine organisms by only surfacing at night, a congregation of marine life forms around their movements up and down through the sea, and the DSL layer ascends and descends with the rhythm of night and day, dark and light. The exact composition of the DSL changes around the world, depending on the biodiversity in each particular area. However, lanternfish, from the family *Myctophidae*, are commonly present within the layer, along with shrimps, squid and siphonophores.

All of which is to say that sound moves in complex ways in the depths of the ocean, but understanding the factors at play helps us to appreciate the entangled beauty of the sea's soundscape, in a way we can only begin to imagine from the sounds we hear at the surface.

Ever since I was a child, I have heard the sea in some way or another. The salt wind that wrapped itself around my life from infancy; the distant beat of the waves that lulled me to sleep; the haunting calls of seals that echoed around the caves and up from the pebble beaches as I walked the coastal path along the cliffs; the throaty caw of fire crow marking her place on the edges, on the fringe between land and sea. When I first sailed the North Sea, I learned to listen to the wind in the sails, the hum of the rigging, the creaking of an old wooden boat. At nights, off watch, I would press my ear against the hull to listen to the water running past beneath. I've always liked dimly hearing the comings and goings of a crew around me, the knowledge that these sounds mean things are being taken care of while I sleep. Sailing on *Balaena*, there was a point where the sea seemed as if it was boiling, the globed heads of pilot whales breaking the surface like bubbles. We could hear their bright whistling sounds swooping through the fibres of the hull. I've heard the whistles and clicks of hundreds of dolphins that have raced and splashed over to the boat to hurtle past and bow ride, keeping time with us. I've heard the long, strange moans of baleen whales, and the long blow of fin whales surfacing. I've been mesmerised by the metronomic click, click of sperm whales searching the depths, speeding up to a purr as they locate their prey. These have mingled together in what has become my personal soundscape of the sea.

There is one whale that sets itself apart with regard to sound. The humpback, *Megaptera novaeangliae.* I first saw a humpback when I was aboard *Balaena*, mere hours out of the harbour, as if the boat's name had charmed the baleen whale to the surface. It wasn't the best sighting – a dark back, a small, stubby dorsal, the humped rounding that gave the creature its common name – but as it rolled and dived, its fluke flicked upwards, displaying the white patterned underside that is an identification feature of the animal, I was overjoyed. The sightings got better. One gold-drenched evening the sun was setting, I was coming out of the galley after washing up, my hands salty, puckered and cold. (All of the dish-washing on *Balaena* is done in salt water during a long sail. Although in the offshore world there is water all around, fresh water is a precious resource.) As I came up through the companionway, all the crew were looking over the guard rail on the port side of the boat. I craned my head around the spray hood to see what they were watching, the water streaked with the setting sun, and there, barely metres from the hull, a humpback whale erupted through the skin of the sea, stealing my breath as it rose out of the water, into the air, until only the fluke remained submerged. Nobbled head, droplets flying like jewels, down the rorqual lines of the creature's throat. The long white pectoral fins moved elegantly through the air, and, for a moment, it seemed that it would defy gravity entirely and continue to travel upwards. Eventually, gravity called the great creature back to the sea, and it came crashing to the surface with a loud splash, the water rippling outwards. I couldn't have planned my exit from the galley more perfectly, and was left feeling incredibly lucky that the whale had chosen that exact moment to breach. It was a little piece of the sublime – one minute with hands in a bowl of cold, salty suds, the next watching this whale – the mundane made magical, but such is the nature of sailing with whales.

We don't fully know why whales breach. Their leaps skywards could be a signal to other whales in the area. Or perhaps they want a glance at a curiosity on the surface. It could be a method of trying to dislodge parasites. It could be simply an expression of joy. Breaching is categorised as one of the ways which humpback whales create sound in the marine environment. We call the breach a percussive sound, along with flipper slaps and tail lobs – drumbeats on the surface of the sea. Their other sounds all happen below the surface, in the form of calls, long moans that resonate through the water, and, famously, their song.

Humpback-whale song has become synonymous with the song of the sea: the swoops, the grunts, the deep soaring rises. The song of the humpback is complex and constantly changing over time. It is made up of themes, and each theme is constructed from phrases. The individual phrases last anywhere between twenty and forty seconds, and can be repeated to make up the theme. The themes are then sung in a particular order, and the song can last up to thirty minutes from start to finish. After which, the whale will pause, and begin again. The song is low frequency, ranging between 30 Hz and 8 kHz, and can travel through the water for thousands of kilometres. We don't yet understand all the reasons why humpback whales sing. Equally, I couldn't tell you all of the reasons why humans sing. I don't know why I am compelled to paint and write, except by some kind of compulsion that eases chaos, and allows moments to pass or be explored with greater clarity. It is something I need, as I need water, food, sunlight and the ocean.

Humpback whales inhabit all of the world's ocean basins, and are migratory creatures, spending summers in the colder, richer waters of the high latitudes to feed. In the autumn and winter, they travel to warmer, tropical waters to breed and calf. The length of these migrations vary between whales. Humpbacks that spend the summer in the Gulf of Maine have been recorded travelling 4,600 km and back, between their feeding grounds and breeding grounds in the West Indies. Humpback whales who feed in Antarctica have migrated to the Pacific coast of Colombia in order to breed, a cross-equatorial migration of 16,000 km there and back. Mitochondrial DNA haplotypes show that contact and mixing between humpback whales from the North Atlantic and the North Pacific has occurred since the formation of the isthmus at Panama cut off the Atlantic from the Pacific. The link between the two oceans has been reinstated by humans via the Panama Canal, but for the whales the route to connect the North Atlantic and the North Pacific is via the historically icy waters of the North-West Passage, which are now alarmingly more free-flowing.

It was previously believed that, although humpback whales vocalise all year round as a means of communication, their song was limited to the breeding season, and perhaps designed to attract a mate. As is the case with many songbirds, it is only the male humpbacks who sing, their complex song ringing out through the water. The song could give the female an indication of the size of the male before she sees him, based on how long he can sing for before he has to surface to draw breath. It is unlikely that the song is a display of fitness based on the animal being able to take the time to sing rather than forage, as many humpback whales will actually fast during their breeding period, relying on reserves built up during the spring and summer. The song of the humpback is constantly changing over time, and also varies with location. The songs of breeding humpbacks in the North Atlantic, North and South Pacific and the Indian Ocean are distinctly different from each other, and yet they share similarities in structure, if not sound. Each breeding season starts with last year's song being sung amongst the whales in that breeding region, and then individual males will innovate, creating a new song. On hearing the novel song, all of the other males in the region begin to adapt, slowly adopting phrases, until, by the end of that season, all of the males in the area are singing the new song. The process repeats the following year, and so the song changes, with an innovative song circulating through the population until it becomes ubiquitous.

It is easy to draw the conclusion that if male humpbacks only sing during the breeding season, their song must be sung to attract females and gain access to mating with them. However, humpback whales have been found to sing during both the migratory period and the period of foraging at high latitudes. Males that have been recorded singing while migrating have been found to swim at slower speeds than males who are not singing, indicating an energetic cost involved in singing along the migratory route. So why do they do it? Why do some males innovate, and others follow the trend? Is it an indication of cultural trends, or creativity in a select few? We know that sperm whales express culture to a degree, so why not humpback whales? What information is being conveyed? What does this say about the nature of whale societies and intelligence? We have so many questions still to answer about whale song, but its mysteries only seem to add potency to the response it provokes in us: we hear in their song an echo of our own humanity. The public only gained access to recordings of humpback whales in the 1970s, when a bioacoustician named Roger Payne released an album of humpback song. This led to a huge boost in public support for the movement to end commercial whaling. It seems humpback song has in fact made us more human.

While we find majesty in the ability of the humpback whale to project its voice over thousands of kilometres of sea, we forget that each of us has a powerful acoustic footprint. The world's oceans physically connect all of the landmasses on our planet, and Western society has a long history of global trade. This trade was once achieved via sail, harnessing the wind to move goods and people around the world, but since the industrial revolution, our lives have sped up drastically. Now cargo ships are powered by fuel. If you have a smart phone, pause your reading for a minute and download the app Marine Traffic. Within seconds, you will be able to see and follow the boats that are currently moving around the world. The sheer volume is overwhelming. Marine Traffic will not show all of the vessels; many smaller private boats, like *Brave*, may not be fitted with an AIS transponder, and therefore will not register. Marine Traffic also doesn't give you a sense of how much noise these boats generate. The chug of ferries, the roar of cargo ships, the engines of private boats. Every time we use a motor to access the sea, every time we order something with the click of the button from half a world away to be shipped via sea, every time we eat food that is out of season or not grown in our country, we add sound to the marine environment. And there are human-made noises even louder or more intrusive than shipping. I've heard seismic blasts as the seabed is searched for oil, and the rising whistle of military sonar as various navies go about their operations – it is so close to dolphin sounds but for some strange quality that is unmistakably mechanical. The impact on wildlife is widespread.

Humpback whales are particularly vulnerable to this sound pollution, as it drowns out their communications. For auditory communication to be effective, it has to be heard. The lee shore of the island of Maui, in the Hawaiian chain, is an important breeding ground for humpback whales. An increase in boat traffic in the area has resulted in mothers and calves moving further offshore to avoid the vessel noise which masks the vocal communications between them. A study in Glacier Bay in south-eastern Alaska, a feeding ground of humpbacks, reported an ambient noise level in the bay of 96 dB. Contributors to this noise were divided into three categories: the sound from boat tours and cruise ships; the roaring of harbour seals, *Phoca vitulina*, which is a seasonal phenomenon; and the natural sounds of weather events. In human terms, 96 dB equates to standing next to a running motorcycle engine; according to the US Centers for Disease Control and Prevention (CDC), fifty minutes or more of exposure could cause damage to human hearing. The study found that the humpback whales in Glacier Bay increased the amplitude of their vocalisations in response to the rise in ambient noise. However, they were only able to do this up to a certain threshold. Not only did the humpbacks have to call louder in order to prevent their vocalisations from being masked, in some cases they stopped calling altogether. Humpbacks in feeding grounds often co-ordinate their feeding effort, using vocalisations, their long pectoral fins, and bubble nets to herd fish that they can then gulp as a group. If they can't hear each other, or if they feel that sound levels are too loud to try and communicate, these shared hunting techniques are affected and the whales can become under-resourced ahead of the breeding season.

Altering the amplitude of the call is one strategy to avoid masking. Other vocal adaptations include altering the frequency over which the call is made, calling more or less often, and switching from vocal to non-vocal means of communication. These adaptations to an increasingly noisy environment have been observed in North Atlantic right whales and beluga as well as in humpback whales. Producing a louder sound requires more energy and, as demonstrated by the study in Glacier Bay, is likely to be achievable only up to a certain threshold. Visual display is only possible if whales are within sight of each other, which is vastly less effective than the far-reaching whale song. Altering the frequency of calls changes the ability of the sound to travel through the water. Ship sounds dominate the lower frequencies, so the whales have to alter their call to a higher frequency, which reduces the distance it travels.

The problems do not stop with the fact that the world's oceans are getting louder. Human activity is changing the very structure of the water at a fundamental level.

Our twenty-first-century lives are reliant on the burning of fossil fuels. Since the industrial revolution, an excess of CO_2 has been released into the atmosphere, and as our lives have sped up, this has increased. The concentration of harmful CO_2 circulating in the atmosphere is actually determined by the oceans, which act as a huge carbon sink, dissolving the gas and keeping it below the surface. When carbon is dissolved in seawater, it becomes an acid. This weak acid then reacts with carbonate anions in the sea and becomes bicarbonate. This process is dependent on the presence of cations, which are added to the seawater through the weathering of rocks, a slow geological process. Biologically, phytoplankton at the surface fix the dissolved carbon through photosynthesis. Some planktonic species also use the carbon to form calcium-carbonate shells. These species either die and sink to the bottom, transporting carbon to the deep ocean, or are eaten and pass the carbon through the food chain. This is one of the reasons why baleen whales are seen as sequesters of carbon, due to the copious amounts of plankton they consume. CO_2 is more soluble in colder, saltier water than in warm fresh water. As a result of ocean circulation and the thermocline, this cold, carbon-rich water sinks, becoming inaccessible for gas exchange at the surface for hundreds of years. This process has been stable for around 400 million years, facilitating life as we know it on our planet. However, over the last 260 years, the excess of CO_2 in the atmosphere has begun to upset the balance. The oceans can no longer absorb or sequester the amount of carbon dioxide we are generating. Not only is the planet warming, and the sea level rising, but the oceans are becoming more acidic.

For many millennia, the pH of our oceans remained stable, alkaline, averaging around 8.1 across the globe, but that number is beginning to fall. Climate models estimate this descent towards acidity at different rates. Two models reporting worst-case scenarios predict a fall of 0.6 pH in the next 300 years, or 0.7 pH by the end of the century. The best-case scenario predicts a fall of 0.4 pH in the next century. Best or worst, the science is unequivocal: the pH is falling, and ocean acidification is occurring. This is affecting the life cycles of countless organisms and altering the ecosystem in ways that we have yet to fully understand.

Ocean acidification also affects the way sound travels. As the pH falls, so does the sound absorption of the particles that make up the water. A fall of 0.3 pH reduces the sound absorption by 40 per cent. As a result, low-frequency sounds are able to travel far further. Although this means humpback calls will propagate over greater distances, so will the anthropogenic sounds that have the ability to mask them.

Human sounds are everywhere in the sea. And the irony is that some of those sounds come from boats that are studying whale sounds. This is why it is so important that vessels such as *Song of the Whale* and *Balaena* are sailing boats, harnessing the power of the wind as much as is possible. Although there is still sound generated under sail as the boat cuts through the water, it is far from the thrum of an engine. Every sailing crew lives for the times when there is no rush to get anywhere, when the engine can stay cool and you can move with the wind. Under sail, a boat comes alive. I stood on the foredeck of *Song of the Whale*, the white of the genoa and staysail curved with the wind, against the startlingly blue sky, in the heat of the sun, and my heart was filled.

The days aboard *Song of the Whale* in the Mediterranean were both exhilarating and challenging. Although this Libyan stretch of the survey was not particularly rich in cetacean sightings, it was still thrilling to be in a sea that was new to me, on a new boat, with a new crew. My position was more demanding than I'd had before, and even though I liked that element of it, I was still in the early stages of my recovery. I worked hard to fulfil my role as well as I could, but I didn't feel as physically strong as I used to be, and my confidence had taken a knock too. We heard more on the hydrophone than we saw, the weather often too windy for great sightings.

It wasn't until the following year that I had a truly singular sound encounter on board. *Song of the Whale* was working with a film crew who were putting together a documentary series on whale culture, 'Secrets of the Whales', for *National Geographic*. On board were wildlife photographer Brian Skerry, who had pitched the series, and Dr Shane Gero, the scientific advisor. They were filming in the Azores, to explore cultural differences between sperm whales in Dominica, the Azores and Sri Lanka, and I was working as crew. The sea here is entirely different. During the Mediterranean survey, I had seen a sea that was facing heavy pressures from fishing, recreational boats and cargo ships, all within an enclosed basin. The Azores archipelago is largely devoid of shipping, and it's too deep almost everywhere to anchor; the volcanic islands rise from depths of 2,000 metres, so deep you don't really need to worry about sailing into obstructions. Occasionally you see the lights of the local fishermen on their jiggers, catching squid. The abundance of sperm whales, especially mothers and calves, in this area is astonishing. Through the night, we would aim to sail and stay close to groups of whales so that the camera crew would have the opportunity to film them from first light. From the boat, we towed the hydrophone, with headphones and a computer set up on the deck so you could hear and see the vocalisations through software.

That night was dark, inky, thick with the gaps between the stars. There were no other boats around. I had begun to find a comfort in the darkness once again. With the headphones on, gently altering course and speed, but mostly drifting, the ocean became an entirely different space. Instead of navigating by sight, with charts, GPS and a course, I was now navigating by the sound of the whales; the only course to keep was their course, keeping their clicks loud, keeping them close, exploring their movements as they explored the ocean beneath. It felt as if I was experiencing their world with them, a world of sound set to their clicking rhythm.

In the majority of the cetacean studies I have been part of, the survey in Libya included, the focus has been on determining which species of whale, dolphin and porpoise are present in the survey area, through sight and sound, and how many of them there are. If we saw one whale, it was likely that we wouldn't see the same individual again. The aim was to create a snapshot of the population, rather than get to know individual animals. Dr Shane Gero has spent his career working with sperm whales in an entirely different way. He is the Principal Investigator at the Dominica Sperm Whale Research Project. In the sixteen years he has been running the project, he has spent thousands of hours at sea with the whales in the Caribbean. He now recognizes individual whales with ease. In Dominica, the emphasis is not on breadth of knowledge, but on depth, as they begin to piece together an understanding of how these whale communities work. They have seen that sperm whales form stable matrilineal social units (the sperm whales in the Azores also form these units, but have been less well studied). In these social units, the whales begin to produce codas – a group of sounds with a variety of rhythms, different from the clicks of foraging whales. Foraging clicks are regular, metered, and speed up to almost a purr as a sperm whale zones in on its prey; they would sound like click, click, click, whereas a coda could sound like click-clickclick, click. The coda a whale produces will define who the whale spends time with, as they associate with other whales that sound like them. This results in vocal clans forming. Ringing out below me in the Azores, I could hear both of these – the sperm whales foraging for food, and the sperm whales producing these social sounds.

When *Song of the Whale* made port in Tunisia to drop off the Libyan crew and collect three Algerian fisheries workers who would participate in the final leg of the survey, I was due to leave the boat. Instead I made the decision to stay aboard for just a little longer, to see the survey to completion. The next leg of the voyage proved particularly stormy. Leaving Bizerte, the weather was extremely rough, with waves crashing over the top of the breakwater at the harbour entrance, but the boat was both strong and safe. Dolphins cut through the rising green swells as we headed out, like a favourable omen. It was a difficult start for the new crew, all but one unfamiliar with sailing boats, who quickly found their heads in buckets. Before or since, I have never seen people be so seasick, their nausea lasting days. I welcomed the hot energy of the weather with exhilaration as we began our windward slog. Standing on deck on a watch, I would duck into the shelter of the cockpit to avoid the spray from a wave, before turning my face to the wind again when it had passed. Our progress was slow – the boat gained speed with the strength of the wind, only to be stalled by the oncoming waves. I received a call on the radio from a coastguard station on the shore. The guard had noticed our lack of speed and asked if we were in difficulty. The only other vessels around were tankers, which were able to maintain their high speeds, whatever the weather. He had mistaken us for one, rather than reading our identification as a sailing boat, and was bemused by our sluggish progress against the weather conditions.

The weather soothed as we approached Palma, Majorca, our final destination. It was a beautiful morning. The sea was glassy, the sun blushing the air and water as it rose over the Balearic Islands. The port was busy, the end of September, full with superyachts and stunning classic boats. Every man and his dog seemed to be out fishing in small craft at the harbour entrance. As we pulled up to the fuel dock and started filling up, the attendant laughed at me. We had had a problem with the watermaker on board – the desalinator that turns salt water into fresh through a process of compression. As a result, we had been showering in salt water to save fresh water. My hair was mussed with it and bleached with the sun. What's more, while loosening the fuel cap, I had cut my hand on the rough deck, and wiped my bleeding knuckles on my T-shirt. I don't know what it is about

Song of the Whale exactly, but her hard-working crew inevitably seems to end up varying degrees of grubby. My T-shirt also had smears of winch grease, and grime from the inside of the cockpit locker, where I had spent hours helping Richard with the repairs of the starboard wheel steering mechanism. For a second, I was offended, but then I looked down at myself and laughed too.

I neither looked nor felt ready for life back on dry land, let alone back at university for my final year of study, but this was where I had to leave the boat.

After some maintenance, *Song of the Whale* was to sail across the Atlantic, ready to start a survey in the States in the spring. I ached to do the crossing. The idea had started to play in my mind of what it would be like to sail across the Atlantic, all the way to America. I wanted to see the Intracoastal Waterway of the Outer Banks, to sail to Nantucket, New England, and eventually New York. I could stay aboard right now, fulfilling a part of that dream far sooner than I could have ever imagined. But if I were to follow that dream, it would mean giving up on another, one I had committed to first.

As much as I wanted to sail, travel, adventure, to put the stress and worry of the last year thousands of miles in my wake, I also wanted to be a scientist. I wanted my degree. The previous year had given me every opportunity to quit, but I had hung on to my education like clutching at a ledge over a precipice by my fingertips. And I knew I now needed to graduate. I needed to wear the cap and gown, shake hands with a university dignitary, and hold the piece of paper in my hand that meant, 'You did that, you held on.' Besides, I had left *Brave* behind in Pembrokeshire, thinking I was going to Devon for a week to do my Yachtmaster. Now I was in the middle of the Mediterranean almost a month later, and I was late for the new semester.

Waiting at the airport in Majorca, I had to scramble to put

together my life back in England, hurriedly trying to re-establish my place at university. I had sailed *Brave* to Pembrokeshire for my planned dissertation study. However, when I'd agreed to work on *Song of the Whale*, they had offered to provide me with a data set for an alternative project, as there simply was not enough time to do both. Years previously, they had surveyed the west coast of the UK and Ireland, and I was going to use these findings to construct my own habitat models of the distribution of common dolphin. Now *Brave*, who was to be my home in Plymouth, was around 300 miles away, moored in Milford Haven. I needed to move her. The weather was entirely and predictably unhelpful. It was now October, and the autumn storms had started to throw their ferocity at Pembrokeshire, while mornings grew later and evenings darker. I stayed with a friend in Plymouth until there was a break in the weather, then headed back home to pick up the boat and sail her back round the south-west coast.

Leaving Pembrokeshire always has a poignant feeling for me, a tug on the invisible thread that ties me to the place. But aboard *Brave*, I was all the more aware of the equally strong force within me – the pull of the water, the compulsion to wander. I left with friends one gusty evening as night fell, an odd time to begin a passage in the growing dark, but it meant that I was able to pilot around Land's End in the light, and on a favourable tide. After Land's End, the sun was already starting to set behind me, marking the passage of time. We had been sailing for a night and a day as we pushed the slowly turning tide at the Lizard, the light glinting off the glass of the lighthouse. Some months earlier, when I had been sailing this same stretch of coast before heading to Pembrokeshire, testing *Brave*'s capabilities and my own, I had reached this point in true darkness, making slow and uncertain progress. I had counted my breaths, in and out, I had watched the flash of the lighthouse, on and off, as I willed *Brave* forwards. The tide, as ever, had served the ultimate lesson in patience, and, gradually, I had got to where I was going in accordance with its pace.

My mood was sombre, solitary; alone on deck, I felt somewhat like the only person in the world, a small loneliness on the night-dark sea. But even if someone had been on deck with me, in that minute it wouldn't have mattered. My proximity to people did not seem important, the walls I'd been building between myself and others had solidified with my back injury, and I found them hard to drop. Being on the water helped. I felt as if it saw me in all my states, as I did the sea – its calms, its fervour, its light, and its darker realms, in which sight was no use. But through sound, you could once again see. Reflective, the feeling started to be mirrored in the water as the wind eased. A small pod of common dolphin pulled me from the well of my thoughts, splashing as they raced over, to ride alongside the hull of the boat. This sighting, the sound, pulled me together, a reminder of where I was going: back to university to finish my commitment to study these creatures.

For all that this passage had started off windy, the night turned utterly still on the final approach into Plymouth Sound. There wasn't a single other vessel on the water, all the usual boats pulled from their moorings and stored ashore for the winter. The only sounds were those of the sea quietly lapping the shore, of halyards running and canvas falling as we dropped sails, the low sound of our motor into harbour. The lights of the channel into the port flashed to the port side, the comforting sweep of the Eddystone far off to the starboard. The lighthouses, those beacons in the dark, always seemed to say, 'We see you. You are here, now, in time and space.'

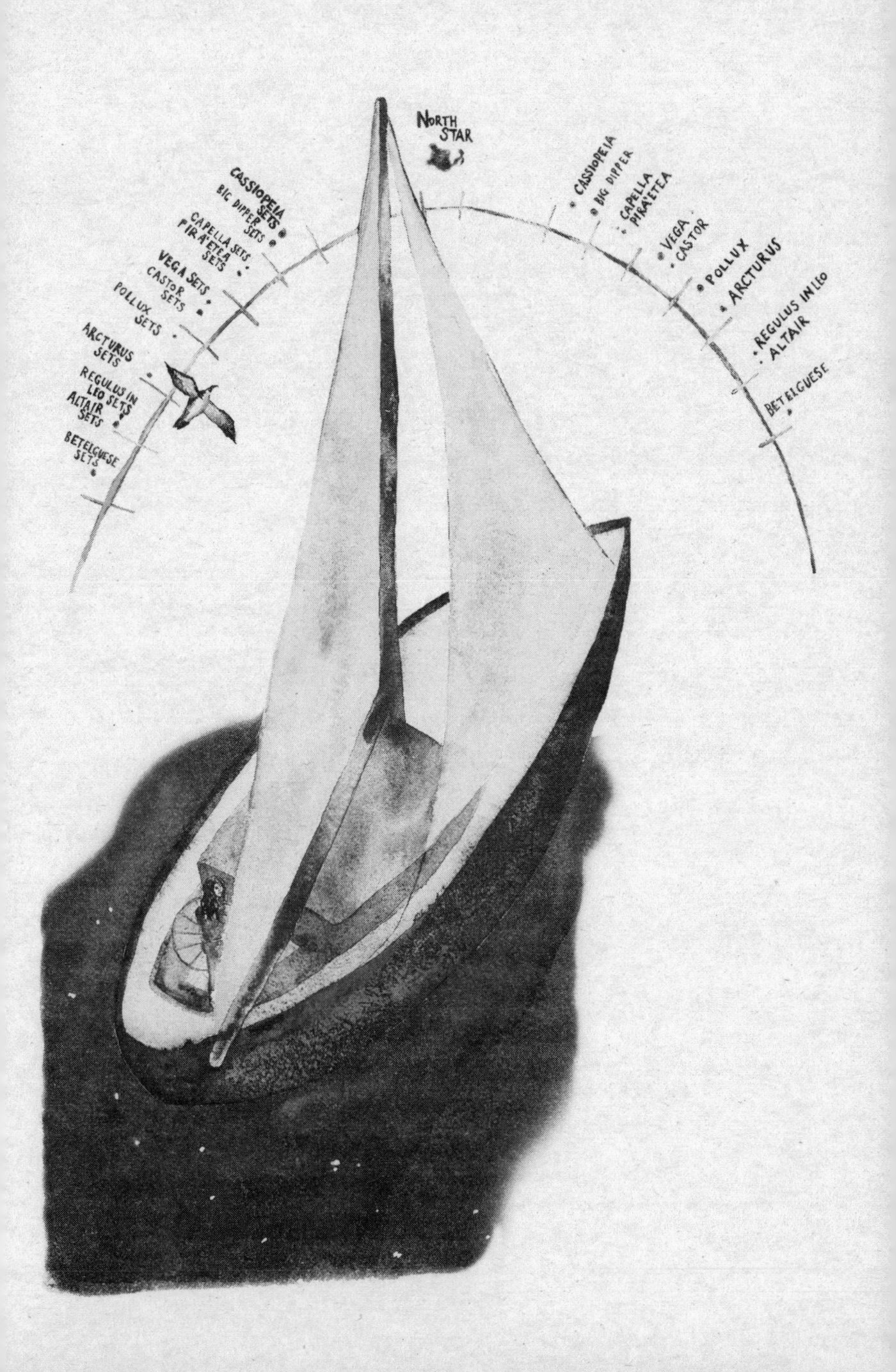

NORTH STAR
CASSIOPEIA SETS
BIG DIPPER SETS
CAPELLA SETS
PIRA'ETEA SETS
VEGA SETS
CASTOR SETS
POLLUX SETS
ARCTURUS SETS
REGULUS IN LEO SETS
ALTAIR SETS
BETELGUESE SETS
CASSIOPEIA
BIG DIPPER
CAPELLA
PIRA'ETEA
VEGA
CASTOR
POLLUX
ARCTURUS
REGULUS IN LEO
ALTAIR
BETELGUESE

6

Shearwater

Beep beep. Beep beep. The sound of my alarm pulls me from somewhere that is not quite in the realms of sleep, nor quite what would be considered wakefulness. Beep beep. Beep beep. As I reach to turn it off, a steady soft purring fills the air. The sound is pure comfort, my cat curled up with his belly against my neck, safe from the cold. The screen of my phone flashes the time with its blue glow: 4 a.m. A rustle as I pull on clothes, and sleepily push my feet into sheepskin boots. The old fibreglass hatch above me creaks in complaint, and the rush of the night air into the small cabin is audible as I heave it open. Best to hurry before I can talk myself out of it. Boots on wood as I climb the companionway steps. The air is still, the water coaxing as it laps against the fibreglass of *Brave*'s hull. A metallic twang as I catch my foot while climbing over the guard rail and on to the pontoon. A snort of breath as my collie dog follows me, her claws clicking in a rhythmic trot. The cat has decided to stay behind, for once the more sensible of us three.

Ice crystals crunch, and the air seems to vibrate with the sense of something other, a lure to those awake to hear the call, a secret to be revealed to those willing to leave the comfort of their beds and go exploring. My breath clouds in front of me, patterns swirling in small eddies. I walk through the boatyard, strange night shadows cast by the boats that have been hauled out for winter; the deep keels that are usually concealed below the surface now tower above me, chocked up on stands. I leave the boatyard behind, cross over the quiet road to the beach. I expect my boots to sink into the sand, to feel the familiar resistance between the grains, but the top layer is frozen, hard and impenetrable. I remove my layers piece by piece, and walk towards the edge of the sea. My collie watches, alert, by the water's edge. She won't come in, unless she loses sight of me for too long.

The surface of the Sound is still, barely a ripple dancing across it. I try to slip in as smoothly as I can, try not to trip on the rocks until I am deep enough that I can give my weight to the water, to start to swim. The sea seems no colder than the air as it wraps itself around me, an inky velvet drape. Then my body registers the chill as my breath grows faster, more audible, my heart rate quickens. Slow, I say, counting as the air passes into my nostrils, counting the exhales as I swim with soft strokes. Slow. It's OK. When I feel that my breath is my own again, I stop swimming, and begin to tread water in the Sound. When this feels easy, I flip my feet up, and lie back, allowing myself to float, held by the water. There, above me, hanging ripe in the sky like a fruit for the picking, the Wolf Moon, with nothing but night sky between her and my skin. The Wolf Moon is the name given to the full moon of January, the first of the year. However, this year, it coincides with a lunar eclipse, staining the full moon red for over an hour. As I float on my back and watch, the colour intensifies further, the blood-tinge deepening. In just over an hour, it will be a blood moon no more. I have positioned myself at exactly the right time, exactly the right place, as the celestial orb calls to the sea around me.

There is something so entirely humbling, something entirely human, about marking one's place in the world in accordance with lunar events. I feel as if I could spill over the edges of myself, as if I could dissolve, slip from my skin and into the spirit of the water. As my thoughts sink deeper, carried on this notion, I dip my head below the surface, no longer floating but swimming down, until the water pushes me back again and I surface into the last hour of night. My collie dog lets out a howl when she can no longer see me, her head reaching towards the moon, throat exposed. This sharp sound rings from the rocks, carries over the water, a disturbance of the peace that is strong enough to pull me from my bewitchment and back to shore. I suddenly feel very aware of exactly how cold the water is on this January night. My collie rushes to great me, dancing around my feet, trying to lick the salt from my legs, barking happily. I pull on my layers, their rough fibres grounding me even as the moon hangs heavy overhead.

I hear the rumblings of an engine, the compression of brakes, then see headlights intruding on the lunar glow. The number 2 bus completes its descent of the hill and pulls on to the road running parallel to the beach. The spell is broken, as thoughts from the city find their way into my head. Although I don't have to catch the bus, I do have a very real exam later this same morning, the first of four in my final year, two this January, two in the summer. I wait for the bus to pull away before retracing my steps back to *Brave*. Once on board again, I make a hot-water bottle and a pot of coffee. I feed both cat and dog, and sit under a blanket at the saloon table. For the next few hours, I write to try and shake my head free of night swims and that haunting lunar glow.

For all my hopes that my final year of study would be smooth, things so far have been rocky. Less than a month after I arrived in Plymouth on *Brave*, I once more found myself staring at the inside of an MRI machine, trying to maintain calm against the pressing claustrophobia and the idea that my newly righted life might be upturned once again. Thankfully, this time, I did not need surgery. I needed an epidural steroid injection, along with medication to soothe the swelling and inflammation in my facet joints and around my lower back. The worst part of the prognosis was that I was put back on the same painkillers that had given me such severe withdrawal symptoms months earlier. I did not want to have to repeat the process, but neither could I function without taking them. The comfort was the promise that this time it would only be temporary, until the injection kicked in.

My mother took me to get the epidural. As I lay on a bed, under a CT scan so the needle in my back could be directed to the most effective point, I felt a horrible pinching sensation that threatened to send me towards a panic attack. There was a strange pressure in my spine, and then a spreading numbness, as if porridge was taking the place of bone and muscle in my left leg. I was not aware until later that a nurse was outside the procedure room talking to my mother, advising her that I should quit sailing, orientate my life towards something more sedentary, less challenging, and avoid ever lifting anything heavier than a small bag. This advice, however well-intended, was contrary to what my consultant had told me. He had always made it clear that he wanted me to be able to get back to my life as much as possible. He encouraged me to sail, and to surf again. Although I still could not bring myself to surf, it felt vital to me to continue to sail, and I feel lucky that I had his words of encouragement to counter the nurse's caution. I have thought about this a lot since, and wondered what advice I would now give my earlier self. That I could still do what I loved, but that maybe I did not have to push so hard. That I could appreciate the small moments of progression without constantly looking ahead, that 'healthy' and 'happy' are small everyday things, with powerful outcomes. That I could ask for support. I learned these lessons the hard way, but I was doing the best I could with what I had.

The following weeks were flavoured with struggle. I went back to *Brave* as soon as I could, anxious not to miss so many lectures this semester. Sitting was deeply uncomfortable, walking equally hard, my left leg felt very weak. With the enforced stillness, however, the overwhelming waves of anxiety were coming back, the abyssal dark once again creeping in. I found people increasingly difficult. I drifted away from many, and was drawn towards the few who brought light into my life. As always, I found solace in the sea. In between time in the library and sitting painfully in lectures, I made my way to the water. Just entering its cold embrace was enough to bring my racing feelings to rest. I would walk to the sea on shaky legs, with an unsettled mind, to come back stronger, carrying with me a small calm. But after a while, my body would begin to ache again and my mind to churn.

I began to think about life after graduation. I saw that the university had a new Master's degree in Marine Conservation, so I applied immediately and was surprised to be offered an unconditional place. This took some of the pressure off my final exams, and I was excited by the prospect of continuing to follow my research passions. But my dream of sailing *Brave* across the Atlantic was also taking clearer shape in my head. I was planning on doing the first part of the journey, from Plymouth, across Biscay, around the coast of Galicia and to the Canaries, with friends, but I now wanted to make the crossing to the Caribbean alone. With all of these future aspirations, and the painful chaos of the past year, and with all of my ambition, it was challenging to settle. For so long I had felt storm-tossed, washed up on the pain that followed from being thrown by a wave. Now I found myself drawing my energy and momentum from the chaos of the tempest, a chaos that, once called upon, I found I could not let go of so easily. Which is perhaps why leaving my cabin aboard *Brave* on a cold January night, sliding through the skin of the water under the Wolf Moon, before the morning of an exam, seemed to me like the only sensible way to be living.

A few months later, I had a chance to stretch my wings. I received an offer to work again on *Song of the Whale*, crewing on their latest survey in the United States. I said yes immediately. Although it meant I would have to submit my dissertation and coursework early, I would not actually miss any lectures, as they thinned out towards Easter, and, besides, I would be doing the science I loved out in the real world, and saving money for my trip on *Brave*.

We were surveying off the North Carolina coast, transiting in and out of the Gulf Stream, and in the waters off Cape Hatteras. My days revolved around taking my watches as a member of the crew. The first hour was observation, standing on a raised platform at the stern of the boat, scanning the water off the port side for cetaceans. The next hour was spent at the helm, keeping the course and writing the ship's log. This was followed by an hour as a data logger, using the waterproof computer in the cockpit to record any cetacean sightings made by fellow observers. The final hour was spent on the observation platform once again, this time on the starboard side. These observation hours were my favourite. I would stand there, swathes of sea around me, my mind settling into the movement of the boat and the waves, always adjusting to keep myself steady. My eyes would methodically search the water, waiting for a flash of movement – the roll of a back, a fluke, a blow. Sometimes there would be a conversation with the observer on the opposite side, the two of us facing away from each other, out to sea, as we swapped stories, or jokes. At other times I would be entirely quiet, using the hour to gather my thoughts, to savour the experience.

Once my official watch was over, I would check the constant jobs list for the boat, and busy myself with whatever needed doing. On a boat like *Song of the Whale*, the list was hugely varied. Sometimes I would be cleaning the heads, the boat toilets, or carrying out engine checks, or maintenance to the onboard generators. I would whip or splice rope, check the rigging, inspect the safety kit on board, or apply coats of paint where needed. After these duties, I would have some time to myself before the next watch. Although I often longed to go to my bunk and read, I would head to the saloon to work on my university assignments.

From the deck, you step down the companionway and into the navigation area. From here, you turn to head aft, and descend a second set of steps, deeper into the boat. The port side is taken up by the galley – space to cook, fridges, spice cupboards and a hot oven. The starboard side hosts a large oak table, which I would often rub with linseed oil. The table is surrounded by seats with cushions that all lift off to reveal an eclectic mix of tinned, jarred and dried foods sourced from all around the world, mementos of whatever country the boat has been stocked in. Between the heat of the oven and the summer weather, it was sweltering in there. The wind whipped outside, and, instead of bringing cool relief, it churned up the water. I would sit, chasing espressos to keep my concentration, my laptop taped down to the table to secure it in place, working away as best I could, as *Song* surged through the sea. My forehead would bead with sweat, my mind desperately trying to keep focus. The unusual conditions in which I worked, my growing tiredness, the lack of internet, were often offset by the wealth of experts I could question as they dropped in and out of the galley for cups of tea, and by the fieldwork we were doing in real time that surrounded me. If I was quick, I would complete enough work to nap, before the whole routine began again. On the days we made port, I would submit my course work, and, once all of that was done, I turned my attention to revising for my final exams.

I had to leave *Song of the Whale* to go back to Plymouth to take them. *Brave* was still safely tied up where I had left her. I stepped on board, but soon found I was rattling around like a pea in a can, unable to settle, unused to time alone after being surrounded by crew. By the end of the last survey, I had been craving my own space, longing to be able to meet my friends for a drink in the warm summer evenings, to smell land smells again – grass and gorse and earth. Now that I was back, I didn't know what to do with myself. Coffee, as always, seemed like a reasonable place to start, so, on the morning of my first exam, I found myself in a familiar café, writing to try and get my thoughts in order. I felt as if I was starting to live parallel lives, that one was blurring into the other until it made less sense and seemed less real. Only a few days ago, I had been sailing on *Song of the Whale* in America, in the waters of the Chesapeake

Bay, while ocean sunfish waved their floppy, loping fins into the sky. I had watched a humpback in the same waters, its graceful long pectoral fins dancing in both water and air. It had been surrounded by the floats of fishing nets, and I was so worried it would be tangled. It seemed to dodge the net artfully, eventually moving on from the area as we sailed away. We had followed a sperm whale with half a tail for a day, after placing a tag on its back. I wondered if it had half a fluke from a birth defect, or whether it had been hit by a ship, a fate that unfortunately is far too common for these whales. We had called him Half Nelson, and I had drawn him in my field journal, to keep his memory with me. Watching the sunset, standing under the blazing stars, feeling the wind on my face both night and day, had been a given. Now, here I was, in a grey city with a great cappuccino. A twenty-four-year-old, with party invites to celebrate the end of university, friends to see, exams to take, and yet this terrestrial life felt insubstantial. I wrote the essays, went to the parties, celebrated with my friends, drank the rum, but part of me still felt at sea.

I was back on shore for so little time, I didn't manage to adjust. The days raced by, and, almost before I knew it, I was stepping out of a cab and on to the docks of Horta, on Faial, in the Azores archipelago. Something in the air felt instantly familiar, like the voice of an old friend calling my name. Both the marina and the harbour were full, boats moored alongside, and swinging at anchor. There was every kind of sailing boat you can imagine, as well as a few bleary-eyed and triumphant crews who had just crossed the Atlantic to reach the island. I saw *Song of the Whale* straight away, with her large blue hull, sturdy build, the observation platform on her aft deck, her tall mast. This time there was no scientific work to do; we were here to host the film crew and Dr Gero for 'Secrets of the Whales'. I was looking forward to seeing a part of the creative process, how they would tell stories about the whales and the environment which is so familiar to me. I sent a message to say I had arrived, and sat on the wall to wait for a ride out in the dinghy. The whole of the dock was painted in a beautiful patchwork of artwork and emblems that told the stories of the crews and vessels that had passed through, a sea story of sorts. The guard rails of boats and the sheets of sails were being used to string up freshly washed laundry, garments flying like flags, and, as the wind blew, the clinking of rigging and boats moving on their moorings filled the air, along with the splashes of colour, all contributing to the celebratory festival feel that presided here. I let the wind wrap around me, knowing that the warm, wet south-westerly would blow northwards, all the way to the shores of Pembrokeshire, connecting me to the place. Winds blow from, tides flow to – an old lesson.

I soon found myself back on deck, alone on a solo watch. There were only three of us on board, and we were moving the boat between islands, where the rest of the crew would join us. While sailing past the volcanic island of Pico, the seabed stretching up into the sky, I took a minute to breathe in my reality. Here I was on a boat, in the Azores, with a flat sea and a soft setting sun. I had done it: I had just finished my degree. Although I hadn't actually graduated yet, I felt sure I had passed. I found myself reminiscing about the last survey days on *Balaena*, in the Gully off Nova Scotia, with the northern bottlenose whales, when university was ahead of me. I had been worried that I wouldn't relate to people, that I wouldn't fit in, that I wouldn't be smart enough, that passion wasn't enough, that I would be found out for not being a real scientist. My fears had proved to be so far from the real challenges I had ended up facing. Now it was over, I was sailing out the other side, and the world seemed to be wide open in front of me. I took my musings to my bunk as

I was relieved from my watch, until some time just before dawn.

When I rose next, it was as if we were floating on glass, such was the stillness. There was a sense of calm, an ease of breath to the beginnings of the day. The passage to our destination in São Miguel would take us right past a seamount in the Dom João de Castro Bank, a steep pinnacle of rock just 13.8 metres below the surface, in waters otherwise 1,000 metres deep. Seamounts are often areas of ecological interest, their formations causing upwellings and concentrations of nutrients around them, followed by aggregations of prey. Maybe there would be whales. I wanted to go and make my morning tea, but anticipation held me on the deck. As we approached the seamount, the water started to reveal the changes in topography: telltale ripples warned of a rapid shallowing. It wasn't whales that awaited us at the seamount, though, but something else. An armada of Portuguese man o'war, hundreds of them, if not thousands, their bladders stretched and puckered as if with stitches, filled with carbon monoxide. They were sails of their own. Flashes of pink warning, the blue tentacles of the siphonophore trailing, armed and ready to sting. They look like jellyfish, like single organisms. In fact, the man o'war is a colony, a multitude of organisms working in a symphony that make up the creature. It was eerie, as if we were sailing into a battle that they had set for us, one single man o'war leading the charge like a flagship. I, alone on deck, they, in the water, as our boat cut through their ranks.

It wasn't until later in the day that we would get whales. Again, I had just woken, this time from a nap. I came on deck in time to catch the fluke of a sperm whale as it dived. There was another, logging on the surface, and, moments later, her calf charged over to her to feed. The calf was surfacing and diving around the mother in a playful pattern, potentially feeding. As I stood there in the sun, watching this incredible connection between mother and calf in wild ocean animals, I was very aware of the feeling of my feet on the deck, the sun on my skin. In that moment, I felt something I had never really felt except on the Pembrokeshire shores. I felt a sense of home. This wasn't the homecoming of an epic adventurer returning to the island of Ithaca, but nonetheless I knew that I had found my way home, to myself. It did not matter what was to come next. What I had done had been enough.

The realisation of this successful navigation struck me. Navigating my emotions had not played a large part in my education, but, over the past year, I had felt newly aware of the opposing currents that flow through me, and of the need to steer a steady course. There is a part of my nature that lends itself well to sailing, which is thinking ahead, outside of the present moment. Although this is very useful for preparing for a big trip such as the voyage I was planning on *Brave*, for provisioning, it can become challenging when the focus gets locked on the achievement rather than the process. Or when looking ahead prevents you from truly noticing the present.

For physical navigation, it is normal to rely on GPS when sailing. GPS is an innovation that makes things far safer, and easier, than they used to be. But I had enjoyed my experience during the Yachtmaster when I navigated the coastal waters using depth contours, fixes and things I could actually see around me, such as buoyage and marks on the shore. This sparked my curiosity about other, more intuitive and observational ways of getting from *a* to *b*. I had begun to learn celestial navigation: how to find my way using the sun and stars.

The European method of celestial navigation is complicated, and was not coming naturally to me. It requires a sextant to take sights, precise time-keeping, mathematics, mental constructs and a series of altitude-correction, arc-to-time and declination tables. You also have to be able to recognise and identify the celestial bodies. The sun is easy, familiar to us all. The planets are brighter than the stars, but you still have to know which planet it is that you are looking at. Becoming acquainted with the night sky in each part of the world, and with each season, takes time – time that I was trying to enjoy and use slowly rather than constantly feeling rushed. The ideas that underpin this method of navigation can seem abstract, for example, the concept of the zenith. To find your zenith, you imagine a line, a thread, running from the very centre of the earth, through the crown of your head, and out into the firmament above. Any heavenly body this line connects with would be at your zenith.

This type of navigation is much easier if we allow ourselves to forget something we know to be true, and instead conjure a simplified world for ourselves – if you can ignore the Copernican truth that the earth is orbiting the sun, and instead imagine a geocentric model where the earth is still and the sun and stars are revolving around us. I was finding it extremely difficult. Staring at the stars I could do. I loved to learn their names, let them ring through my mind, to be recited quietly on night passages as I tried to connect the constellations with invisible lines. But gazing at the stars and finding your way by them are entirely different things. There are so many possibilities for error, between life on board, disrupted sleep cycles, problems with time-keeping, as well as the weather. For all that I struggled with applying the method, I frequently found myself marvelling at the mathematics behind the altitude-correction tables I was learning to use.

The altitude-correction table is necessary because the mathematics that underpin the navigation are based on the sight with the sextant being taken at the exact centre of the celestial object, with the observer standing at the very centre of the earth.

In practice, of course, this is impossible, so corrections are given. My particular favourite is the altitude-correction table for Venus, largely computed by Maria Mitchell. A nineteenth-century astronomer, a mathematician and a naturalist, Maria Mitchell worked at the US Hydrographic Office for a part of her life. She was not a doctor of astrophysics, astronomy or mathematics. Although she was awarded a professorship at Vassar College, she did not have a college education herself. She was born on the island of Nantucket and had fostered a fascination with the stars from an early age. She was a young girl filled with as much curiosity as I had had, but with a far better head for numbers. While I had been rummaging in tide pools, exploring cliffs and caves, swimming in the sea and staring at the horizon, the young Maria Mitchell had been watching the skies. When she was only twenty-nine years old, she discovered a comet, which came to be known as 'Miss Mitchell's Comet' (now called C/1847 TI); for her efforts, she was awarded a gold medal from the King of Denmark. I liked that the table I was using had been produced by such a woman, yet another female guiding light in my life.

My struggles with the sextant had led me to investigate other ideas on navigation. When European explorers 'discovered' the first of the Polynesian islands, the Marquesas, in 1595, they found to their surprise that the islands were already inhabited. This sparked some big questions in European society, which then took hundreds of years, and advances in multiple disciplines of research, to begin to answer. Where had the Polynesians come from? How had they settled the islands in the Pacific, from Hawaii to New Zealand to Easter Island, a region spanning 800,000 square miles known as the Polynesian Triangle? One idea was that the Polynesian islands had previously been a continent that had been flooded, leaving behind only high peaks as habitable land and isolating the populations there. When it became clear that there had been movement of people between the islands, theories were proposed that assumed the Polynesians were able to conduct short, accurate voyages between islands, but that any long-distance voyaging was the result of drifting rather than actively sailing.

Some argued that the Polynesians had come from the east, from the coast of South America; and some thought they had come from the west, from Asia and Micronesia. There were centuries of conflicting hypotheses, strong opinions, archaeological and anthropological discoveries, the development and implementation of carbon dating, geographically modelled scenarios. One of those who weighed into the debate was the explorer Thor Heyerdahl, who believed that the Polynesians had arrived on the Pacific islands from South America, drifting there on the ocean currents, accidental voyagers. To prove this theory, he constructed a raft, the *Kon-Tiki*, in traditional Polynesian fashion, and in 1947 sought to make the drift voyage himself. Heyerdahl corresponded with Rachel Carson, writing to her about the strange bodies of squid he saw in the night, and the flying fish that hurled themselves on the deck of his raft. Although Heyerdahl did successfully reach a Polynesian island, what is often overlooked is the fact that the *Kon-Tiki* had to be towed, under motor, fifty miles out to sea in order to be swept up in the Humboldt current and carried along to meet the South Equatorial. Without this intervention, the raft may have never drifted away from South America.

Thor Heyerdahl and others who adhered to the drift theory failed to take seriously the native tradition of wayfinding. Three decades later, however, the skill of the early Polynesian navigators was in no doubt. In 1976, Mau Piailug, a master wayfinder from Satawal in the Carolinian islands of Micronesia, successfully navigated *Hōkūle'a*, a replica of the early Polynesian voyaging canoes, built by the Polynesian Voyaging Society, from Hawaii to Tahiti. It was a journey of 2,600 miles across a stretch of water that was unknown to Mau. And, since Hawaii lies further to the north than his home island, the starscape would have been unfamiliar too. He had never sailed with this particular crew, which consisted of Hawaiian islanders who had been cut off from their wayfinding heritage. Before departing, Mau briefed them, asking them that, for the duration of the voyage, they should forget about their lives on land, put aside notions of the shore, and let their world become the raft, the sky, the water, the wind. Mau had no traditional compass, no sextant, and no GPS, and yet he found his way. The oceans and the skies spoke to Mau. He read the stars, as – from his perspective – they rose from and fell into the ocean night after night. He read the water itself, searching for familiarities in the swells, observing how the water hit and moved the hull. He used the cloud formations, the birds and the fish to find his way.

These were skills that Mau had been refining his entire life. He had been chosen to be a navigator as a child, and from an early age he was placed in the tide pools of his home island, to learn the feel of the water, to begin to read the messages revealed to those who know how to look. He also studied through an oral tradition, the genealogy and techniques of the wayfinders who came before him that were not recorded in writing. This practice was reverence, akin to religion, a tradition, a responsibility. He belonged to a people who had not accidentally drifted but had skilfully found their way. Those who saw him navigate described his wayfinding ability as almost like a sixth sense. After the first voyage of *Hōkūle'a*, Mau took an apprentice, Nainoa Thompson, who would himself successfully navigate the same passage years later. Thompson talks of Mau's ability as something that was outside the analytical mind, steeped in deep feeling, intuition, learned knowledge and spirit.

When Thompson was learning to navigate as the ancients did, he studied in a planetarium, using a star projector to become entirely familiar with the night sky and the perceived movement of the stars. In time, and as instructed by Mau, he developed a star compass. The star compass is a mental construct, entirely different from the magnetic compass of Western navigation. For the wayfinder, the raft is actually still, and it is the ocean and the stars that move around it, islands being called from the deep. In the star compass, the horizon is split into houses, each house a place where a familiar star rises from or sets into the ocean. What sounds simple enough in theory is, in practice, an incredibly complex exercise, requiring a wealth of knowledge committed to a state where it becomes almost innate knowing, filling the subtle corners of the mind. In the northern hemisphere, Polaris, the brightest star of the Ursa Minor constellation, can be used to find north. From the latitude of Hawaii, Thompson would have been able to use the Southern Cross to find south. Thompson learned to pay attention to the rising and setting points of the sun, rising always in the east, setting in the west. The mid-day meridian passage of the sun, such a cornerstone of the Western principles of celestial navigation using a sextant, was essentially useless, as it does not relate to the rising or setting point.

On nights thick with inky black, Thompson was instructed to use the feel of the waves, the way the water moved beneath the hull to find his way. This too was a part of the star compass. Unlike Mau, who had been learning to follow the tidal flow of water since he was a child, Thompson struggled to tune into this subtle movement. Birds were another element to be added to the compass. Pelagic birds – the petrels, the albatross – who spend most of their lives on the open ocean, were no use in finding land, so the wayfinder looked to the land-based birds as they returned home from foraging trips in the evening, leading the way towards a hard-to-spot atoll. The presence of certain sea mammals in a particular area, or changes in their behaviour, also offered clues in this oceanic puzzle.

This last part I felt I understood. Sperm whales are creatures of the deep, diving to thousands of metres, so I knew not to expect to find them in shallow waters. When I was younger, I had learned about topography from watching the harbour porpoise, *Phocoena phocoena*, feed at each end of Ramsey Sound. If the tide was flooding, I would find them at St David's Head; on the ebb, I would perch at Pen-Dal-Aderyn. Porpoise can be hard to spot, small, dark backs breaking the surface among the overfalls and whirlpools that often form with the strength of the tide as it rushes through the Sound. They are easier to hear – a puff as they surface and gulp the air – as long as the sound is audible above the rush of the water. Their feeding patterns tell a story about the underwater landscape. The Sound is a stretch of water running between Ramsey Island and the mainland, a deep, steep-sided underwater valley, with the pinnacle of Horse Rock jutting just beneath the surface in the middle, and The Bitches reef stretching out like a ghost bridge, haunted by the ships that have been wrecked on its rocks. The tide runs through the Sound in a race, squeezed between the island and the mainland. The powerful flow scours up nutrients as it goes. When it hits the shallowing at either end of the Sound, the water wells upwards, providing a rich feeding ground for fish and, in turn, for porpoise. I would watch them for hours, black backs breaking blue water, the flash of white as gannets circling overhead dived, spear-like, into the water. The tide would start to slacken, and the porpoise would begin their journey from one end of the Sound to the other. They taught me whether the tide was on the ebb or the flood, whether it was nearer to spring or neap, depending on how far from the mainland they were feeding, as the strength of the spring tide usually pushed them further offshore.

Birds, stars, water, cetaceans: all of these were observed and learned by the Polynesian wayfinders, tapping into their patterns and rhythms as a way of interpreting the seascape. Wayfinding is a tradition to be learned and passed on, but we know less about the navigational ability of animals. A male sperm-whale calf, born in the Azores, will navigate his way into colder, higher latitudes, such as the waters I had sailed on *Balaena*. The humpback whales migrate from feeding grounds in Greenland to the Caribbean to breed. Birds navigate their way around the world with impressive ability. The islands of my home in Pembrokeshire are the breeding ground for over half of the world's population of Manx shearwaters, *Puffinus puffinus*. The majority breed in burrows on Skomer and Skokholm, with a smaller population on Ramsey Island. These birds are best seen in their element, skimming over the surface of the water at startling speeds, with an otherworldly elegance. I have spent many hours in a perch up a mast, watching the song of their flight, a dance through the air as they glide over the waves. On a summer's evening in Pembrokeshire, as dusk falls, you will find rivers of these birds streaming home from a day's foraging, returning to their mate or chick in their burrow. Tens of thousands of them fill the sky. Their backs, the top of their wings, a deep black, their underside a bright white. Flashes of yin followed by yang as they flit over the water. They lay their eggs inside burrows on the islands. The incubation for a Manx shearwater egg is shorter than for the wandering albatross, an average of fifty-one days rather than seventy-eight. But, like the albatross, the pair take turns, supporting each other as they incubate the egg and forage to sustain themselves. Once the chick is hatched and is large enough, both adult birds will be out at sea simultaneously, the chick safe in the burrow when they return to feed the fluffy creature as its flight feathers grow. Along with Ffion, my first boss, a woman renowned for rehabilitating injured birds, I once saved a Manx shearwater from being drowned by three black-backed gulls. When I scooped the bird from the water, despite its stressed and bedraggled state, I was struck by how light it was, the air spaces in its delicate bones speaking of riding the sky over the sea. The tiny creature I held in my palm was a great voyager. We took the bird to Ramsey, where it was released later that day.

These birds will think nothing of a daily foraging trip of hundreds of miles before returning to their burrows. Their yearly migration takes them from Pembrokeshire to their wintering grounds, off the coast of Brazil or Argentina, then back to Pembrokeshire to breed again in the very same burrow in the summer. Because their flight is so fascinating, people have been experimenting with the Manx shearwater for many decades. In 1939, a man named David Lack took it upon himself to remove three breeding shearwaters from their burrows on Skokholm, and to transport them via train to Devon. His idea was to then release the birds, and see how long it took them to navigate and fly back to their home from this presumably unknown location. As might have been predicted, the birds, taken from their burrows, from their mate and their egg, and placed in the alien environment of a train carriage, were hugely stressed. Two out of the three died during the journey, never to return to their mate, presumably causing an egg to be abandoned, the breeding season to fail. The surviving bird was released from Start Point, a jutting headland that causes a tidal race and turbulent water that has recently had me helming a boat with my head in a bucket. The same evening, the bird was seen back in its burrow on Skokholm. It is likely that it had arrived earlier in the day, but waited until darkness was falling to come ashore. As gracefully as the Manx shearwater flies over the sea, they become very clumsy on land. When they fly over the water, they glide using updraughts of air that form over the waves. These propel their flight. When they fly over the land, these updraughts are lost, and they can no longer flit and accelerate with the same speed and elegance, so become very vulnerable to predation. If you visit Skomer Island, you will see that not only is the ground broken with multitudes of burrows of both Manx shearwater and puffin, their eggs and chicks tucked up safe underground, but amongst the thrift, bluebells and spongey tufts of grass, there are hundreds of shearwater carcasses. Some are picked clean and bleached by the sun; others still have feathers that blow and move with the wind. These are the birds who never made it to the safety of their subterranean nests.

Similar experiments continued, although with more concern for the welfare of the birds and the impact of their removal on the breeding season. Perhaps the most remarkable was a Manx shearwater taken from Skokholm all the way to Boston, Massachusetts. From Boston, the bird was released, and just twelve days later was found back in its burrow in Pembrokeshire, having made a journey of 3,067 kilometres in less than a fortnight. Although the Manx shearwaters undertake the migration to South America every year, they have time to provision and replenish themselves for the journey, during the period in which they are rearing their chicks. This incredible bird had been forced to make an unplanned journey, via a partially new route, without this prior preparation.

We know that birds such as the Manx shearwater and the wandering albatross are incredible navigators, but we still don't know exactly how they do it. Just as the navigation abilities of

Polynesian wayfinders seemed implausible to early European colonists, so we are still learning how these avians navigate the globe. The current assumption is that their navigational ability lies somewhere between the learned and the innate, drawing on orientational cues from the sun and stars, and an inner awareness of the earth's magnetic field. Perhaps shearwaters are constructing their own unique compass, the compass of the winged wayfinder.

My nights in the Azores were dictated by navigating by the sounds of the sperm whales as they foraged in the deep. Tracking them acoustically, we would stay with them all night to be able to film them as the sun rose. You knew that you were close to the whales if the clicks were loud. If they started to fade, you knew they were moving away. They were our destination, rather than any fixed point on a chart. If they started to fall quiet, you had to make a guess, altering course to either port or starboard. You had to wait a moment for the hydrophone to stream out behind, on this new course. Once it had settled, if the clicks grew louder, it meant you had guessed correctly; quieter, and you had turned away from the whales. Things weren't quite that simple – when the whales are at the surface, breathing after a dive, they often pause their vocalisations while they take in air. In the dark of night, it was an extremely singular experience, but incredibly rewarding if daybreak coincided with the blow or fluke of a sperm whale, or the animal logging next to the boat.

Although the weather was blustery, the filming wrapped successfully. This time, I was free to stay aboard *Song of the Whale* for the next part of the journey. I was no longer splitting my life, and, contentedly, I had nowhere else to be. We set sail for the UK, three of us on board, the water beneath like lapis lazuli made liquid. Streaks of light danced through the surface waters before descending, lost to the dark of the deep. These streaks of light brought to mind the strings of my harp, and I wondered, if I could pluck them, what sound they would play. We were joined by spotted dolphin when leaving the islands, a farewell from the Azores.

The bright skies of our departure swiftly gave way to the brewing of a storm that broke over us that night. I was being tossed around my bunk, tensed against the rocking of the boat, when I heard my crew mate calling my name. Normally, she would have the steering on autopilot, while she came down into the cabin to wake me up for my watch. When I heard her calling, I woke up instantly, as I knew that something must be wrong. You don't ever find a heavy sleep while sailing, a part of your brain constantly assessing the sounds of the boat, wondering whether you are needed. It is part of the skill of it. I quickly layered up against the chill of the night, staggering about as the boat shifted around me, searching for a hat, my boots, waterproofs, a life jacket. When had I last brushed my teeth? No time to think of that now. The lights inside the boat were set to a soft red glow, to help protect night vision, my eyes acclimatising as I made my way from the cabin to the navigation area, and out on to the deck.

The world was wild, the wind roaring around us. Judith stood at the helm, her red hair whipping in that wind, salt-slicked tendrils like flame. She explained that the autopilot wasn't working, that it kept losing course. She had been alone on watch, steering by hand at the helm for over an hour. I asked if she would stay there a minute longer, while I went and made the usual checks that typically solved the problem. I couldn't find anything obviously wrong, so resorted to turning the autopilot off and on again. It seemed to do the trick, and Judith went to get her share of the night's sleep we were rationing among the crew. The boat was flying, over fifty tonnes of metal filled with people, water, food and research equipment, moving through the water at eight knots with barely a scrap of canvas. The idea has always seemed to me perfectly madly marvellous, the way we can travel the seas using the wind so well. The wind was thankfully behind us, the mountainous sea following, rather than something we had to beat into. Although the autopilot had initially held, the problem returned. The boat had come off course, and I was worried about crash gybing, with the stern coming through the wind unexpectedly, or being caught broadside by a wave as the course was lost, so I too was steering by hand.

Helming a course by looking at the compass is a notoriously difficult thing to do. While you are looking down at the spinning instrument in front of you, it is easy to back a sail, or steer in a horribly wiggly line. I always get weirdly confused about which way I need to turn to correct my course. Instead, it is best to take stock of your surroundings, glancing at the compass now and then to check you are still going in the right direction, and find an object sitting on the right bearing. This can be a landmark, a feature on the coast. You can use stable cloud formations or stars. I have a tendency to use the wind, something I can feel. The night was dark, the sea a roiling tempest around me, but I sat the wind just off my right shoulder. My hands were on the wheel, my feet were on the deck, and I could feel the world around me. There was energy everywhere – the wind in the sails, the sea spray that lashed across me. It was hard not to come alive with it. I suddenly realised that my old, borrowed waterproofs were that in name only. The red of the jacket and salopettes was faded to pink, the proofing all but gone. I vowed to buy myself a new set the minute I was back on shore. I took in my situation, and began to laugh at myself, my voice adding to the wild of the wind and water. We were somewhere west-south-west of Biscay, north-east of the Azores. I was alone on deck, at night, in the middle of the storm, clipped to the boat by the lifeline attached to my life jacket. Back when the rain had been pounding on the roof of my cottage bedroom, I had longed to be out in it, fully part of it. Now I stood in the night, and made it my own.

A great wave was growing and growing on the starboard aft quarter of the boat. Just as I was thinking that the wall of water could not possibly get any higher, all of a sudden, the peak split and fell away. Happy that I was not about to receive a thorough soaking, my relief quickly turned to surprise. The night was as full of petrels as of wind. They had been accompanying the boat the whole time, a winged symphony. And then there was another surprise. A bird crashed on to the deck. A shearwater who had been riding up the back of that great cresting wave, and hadn't seen the boat. We looked at each other with a mixture of shock and confusion at its ungainly arrival. I was about to leave the wheel to scoop up the bird and throw it so it could get airborne again in the maelstrom, but the bird shook its feathers, waddled gracelessly to the stern, and plopped off the back, once more into a world of wing and water, as if this was an embarrassing incident it would rather we both forget. I wondered if, like me, the bird was returning to the UK.

As the bird joined the many around me, no longer individually distinguishable, I continued to steer into the storm. I had found the flow of it now and my thoughts began to roam. Mau had asked his voyagers to set aside their lives beyond the boat, to forget about the world beyond the here and now, but I kept thinking about my journey. Instead of a star compass, my mental construct was a chart, one with its own particular set of waypoints. There was the lighthouse of Strumble, whose beam had drawn my thoughts away from the shore and out to sea. There were the harbour porpoise in Ramsey Sound, who had taught me about tide. There was the dead pilot whale that had washed up on the beach and made me want to know where it had been, how it had lived. There was the constellation of Orion, the great hunter, whose stars had been an anchor to me from my early days in the garden of the cottage. My starscape had grown from this one constellation, to include Cassiopeia, Lyra, Corona Borealis, and Polaris, the North Star. Fire crow, a creature who exists on the margin between land and sea, and made me think that I too could find my place there. And then the whales. The sperm whale that had showered me in the night, entering my realm for a moment and reminding me of the things our species share. The clans who had shown me the importance of community, and strengthened my bond with and appreciation of the women in my life. The humpback outside Aberdeen Harbour had been another omen, an affirmation in space and time, as I was starting to study. The bottlenose dolphin who had raced through the surf with me that singularly glorious morning, which had brought a happiness that would sustain me through the hard times to come. The albatross that had flown through my dreams and inspired me to sail again. All of these creatures, these moments when our lives had touched, were waypoints on my journey. But they were also waypoints in my understanding of the oceans, and they showed me how much there is still to discover about the water and the creatures who dwell in it. I knew how to listen to the sounds of the sea, and how important the voices of the ocean were. I knew that I wanted, in some way, to be a voice *for* the oceans.

Due to the strength of the wind and its favourable direction, we flew through the rest of the journey. We made the passage across the Bay of Biscay, and then made port in Lymington, on the south coast of England. For a few weeks, I stayed aboard *Song of the Whale* by myself to work, before Richard and I sailed her to her home port in Ipswich. I had never sailed through the

Dover Strait before, the busiest shipping channel in the world. There were tankers moving goods around the world, and ferries crossing between the UK and the Continent. They made the passage seem like some sort of strange video game, the predictable movements of the large fast ships interspersed with the erratic course of fishing boats to keep you on your toes. We sailed through a night, a day, and a night again, arriving on the Orwell River at dawn. This final section of the journey was sublime, a welcome break after the chaos of the English Channel. A blushed light rising, tree-lined banks, and the smell of the late-summer fields in the air. I left the boat that morning, and took myself straight back to Plymouth, unbelievably tired, for my graduation ceremony.

I wore the hat, the gown, I shook the hand, I received the certificate. I stood with my parents before Smeaton's Tower on the Hoe, the Eddystone out to sea behind us for the photos. The moment meant so much more than the degree – a transformative chapter in my life finally closed. By stars and birds, wave and whale, I had found my way.

7

Barnacle

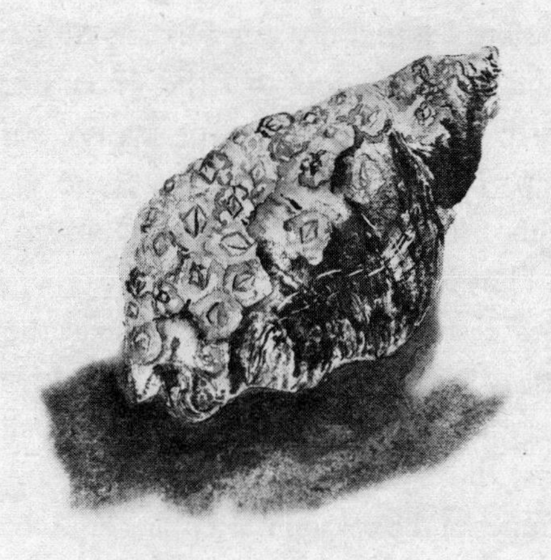

Today I drove my mother's car until the road ended. I pulled up, parked, and started to walk the old familiar path. It has changed in places, moving in from the coast as erosion takes land back to the sea, but my feet still know where to go. My legs are both stronger and wearier than they were when I was a child. There is less of the impossible lightness and energy of youth in my step, but now every time I plant my feet, I feel a connection that has a deeper resonance. The air is bright for December, the sun warm. The light catches on silver cobwebs, thousands of glistening strands as if spun by some kind of thistle witch. There is one tangled in my hair, streaming out behind me, but I cannot bring myself to remove it. The tide is ebbing fast. It is two days before the winter solstice. Somewhere down in South Georgia, a young wanderer chick may be emerging into the world of the nest for the first time. Last night was the last full moon of the year, cold and bright as it hung in the sky.

After following my feet, I sit on an old stone that juts out into the sea, covered with thick lichen, my seat at Pen-Dal-Aderyn, at the south end of Ramsey Sound. For years, this was the edge of my world, before I found my way further. Although today I will stay on land, I already feel that the water surrounds me. Its hushed roar is constant as the river sea flows out towards the Celtic Deep. The patterns in the water are so singular to this channel, this stretch of water between Ramsey Island and the mainland. They are as familiar as the lines that are starting to show on my weather-worn hands. I can read them like the words on this page, an upwelling here, a back eddy there – the language of the water. I turn my head to the left, changing my view from across the Sound to Ramsey, to the stretches of St Brides Bay, Skomer and Skokholm. In doing so, I find that I am being watched by a grey seal. A female, light and silvery, the sun glancing off the slick of her head. She was fishing in the back eddy, between the point where I am sitting and the Shoe Rock, which is just starting to emerge from the water with the fall of the tide, golden kelp meeting the light as it clings to stone by its holdfast. The water is building on the north side of The Bitches reef, damning itself before cascading through the bridge. There are people walking behind me on the coast path, bright coats, map cases and walking poles in hand. I wonder if they love it here like I do, how it feels for them to watch the tide race past.

This is where I come to understand how water can move – a swirling, falling waltz as the tide drops. And this is where I come to feel at peace, utterly understood, a soul reflected in the mirror-glass patches. Even the smooth water hides the tide beneath. I don't live here any more. My collie dog is now too old and tired to make the journey with me. Instead of walking as my shadow on the path, she's tucked up in front of a fire in the cottage. My old ginger cat died the day before I could make it home, and the great ash tree in the garden is brittle with disease, the branches no longer able to hold my weight. Time has moved like this river of tide, and I with it. In the summer between finishing university and actually graduating, I made the decision not to take the place on the Marine Conservation Masters. Although I am for ever grateful for my education, and although I will never forget the words of my Ancient Mariner, I felt it was becoming unhealthy for me to split myself between work and study. My passion remained unchanged, and I had a storm of ideas brewing inside me, but I was beginning to sense that I needed to find my way outside academia. In the meantime, I could work on *Song of the Whale*, saving money for my trip on *Brave*.

There was hardly a minute of time on land when my mind did not drift to the sea, wondering what it would feel like to sail out past Plymouth breakwater on *Brave*, leaving the city, with its good times and bad, behind me. I wondered what conditions I would meet in Biscay, whether there would be the fin whales I associate with the bay, their long blows reaching to the sky like pillars as they come to the surface for air. I wondered what it would feel like to set out from the Canaries, without human company, travelling with just my boat, the stars, the sun, the sea. It felt like something I needed to do, a journey synonymous with healing, a sign that it would close that part of my story as well. I was constantly scouring eBay and boating forums for parts for *Brave*. I had ordered my charts for the trip and was planning my route.

*

In the years since I graduated, the world has changed in ways no one could have imagined. My plans for 2020, like everyone else's, were thrown into disarray. My longed-for solo voyage across the Atlantic on *Brave* was postponed, and then began to seem increasingly unlikely; I began toying with the idea of selling *Brave*. My original plan had always been to put her up for sale once I reached the United States after the crossing. Although she was beautiful, and symbolised freedom after struggle, she also reminded me of a time in my life that had held a lot of pain. *Brave* had always been my space, a space that it was difficult to let anyone else truly be a part of. I associated such a strong sense of independence with her that I had rigged a system to move the heavy life raft off her decks by myself, rather than just asking for help to lift it. Now I wanted to, felt able to, share more of my experiences at sea with my family and more of my friends.

I had seen *Larry* years earlier when I was introduced to her owner by friends in Cornwall. *Larry* was a gaff cutter, built in Dorset in 1907, and had been owned by Sue Singer and her late husband for three decades. They had crossed the Atlantic, taken her north in the same freezing fog I had sailed in on *Balaena*. They had ventured further, to Greenland. They had visited Shetland in her, and the Lofoten Archipelago in Norway. My partner Henry had sailed across the Atlantic a few years previously on a 98-foot schooner of a similar vintage, and one of his crew mates had also owned *Larry*, not once but twice, for after he sold her he couldn't forget about her and bought her again. Between the two of us, we knew around sixty years of *Larry*'s ownership history. She was built to go anywhere, and had been tested time and time again. We could be the next part of her story, as she could be of ours. It felt as if the world was pushing me towards her, rather than a misguided notion of what I thought I needed. We bought her, and made her our own, and we have filled her with love, hope, good food, better friends and the spirit of adventure. Our first extended journey took us from Cornwall along the south coast of England, across the North Sea, with a nod of respect to my Mariner, through the Kiel Canal, and to the Baltic Sea. These waters were new to me, a new feel to learn, drastically different from the wild Pembrokeshire shores. New, fresh – that felt good.

My heart soared when I saw the first fin of a porpoise breaking the surface. There was a small pod, about a mile off. I might not have noticed them if I hadn't known they would be there, but I have spent my life so far training my eye. Suddenly tears began to fall and my vision blurred. They were tears of joy, that the world had changed so much, that I had changed so much, and yet I could still come back to Pembrokeshire, and spot my old friends. They weren't here to see me, of course. They were here to feed, fishing in the flow, but the fact that they keep to such familiar patterns in a world of flux stirred within me the deepest gratitude.

In the spring, we had set sail on *Larry*, travelling eastwards along the English Channel, and north-east across the North Sea. It was just Henry and me on board, a small crew, but our combined experience was enough to make the trip safely. Although the fundamental principles remain the same, sailing *Larry* is entirely different from sailing *Brave*, *Balaena* or *Song of the Whale*. Those vessels are all modern boats, *Brave* being the oldest of the three. They are all Bermudan rigged, with a triangular mainsail running up the mast from tack to head, the clew out on the end of the boom. Both *Song* and *Balaena* had a staysail up forwards, and all three a genoa on a furler, to be wound around the foil when not in use, and unfurled to fly with the wind. On each of these boats, you use winches to hoist and trim sails. But *Larry* is gaff-rigged. Her mainsail is four-cornered, an additional wooden spar, the gaff, running along the head of the sail. To hoist this heavy gaff, instead of wrapping a single halyard around a winch, you must sweat and tail both the peak and throat halyards up in combination. The throat attaches to the end of the gaff closest to the mast, the peak to the aft end. To sweat, you grasp the halyard with both hands, allow your body to be somewhat fluid, and, with a motion that becomes smooth with practice, pull out on the rope and then down. When tailing, you take up the slack generated with this movement, pulling in all you can on this line, secured with some purchase around a pin. All of her canvas is heavy, cut and made in a traditional style, and shines white in the sun, save for her topsail. This canvas is dyed red and requires its own mast – the topmast, which extends from the last few feet of the main mast up into the air.

When we hoisted her sails for the first time, and felt *Larry* move with the wind as we travelled out through the Carrick Roads, the sun was bright, the wind fair and following. Although I had seen myself at the helm of *Brave*, leaving Plymouth, this was somehow more poignant. Now I was at the helm of a historic boat, adding my name to her history. I was sailing out through the waters where I had first come to realise how seriously I had hurt my back.

Henry and I took it in turns to sail, two hours on deck, two hours asleep, the usual order of day and night replaced by this rhythm as we sailed through both sun and starlight, supporting each other. Even tucked up in our bunk, you can still see out through the companionway, out into the cockpit, so we could easily help each other if problems arose. To the left, a brass porthole, through which you glimpse a circle of sky and sea. In the calm weather, I would perch on the back of the cockpit combings as I was helming, taking in the creak of the wood and the rigging as the boat moved through the water. *Larry* was made from trees. They had been felled over a hundred years ago, sawn into frames and bent into planks. Although many of these had been replaced and repaired over time, some were still original. Every part of her was designed to be slightly flexible and respond to the water. She needs the sea, the salt water swelling the planks of her hull, the planks of her deck, so that they keep the water out. Sitting there in the sun, I was letting my ear adjust to her sounds, learning what was typical for the boat, so I could identify when things went awry. For now, she carried us safely over the water.

The crossing over the North Sea was smooth, the sea undulating with a regular roll. Although the spring nights were cold, and we shivered in layers of wool and fleece, they were pleasantly short. And each day, the light grew longer. Our first port of call after a week was Bremerhaven, on the River Weser, in Germany.

We stayed there for over a month, fixing things on board, and touching up layers of varnish until the boat shone. As spring rolled into summer, the weather turned hot and humid. The pressure would build throughout the day, both in the air and in my head, my forehead pounding. When I thought it could get no worse, cracks of thunder would peel above me, before the skies opened in a deluge. This pattern seemed to repeat, day after day, until it was time for us to move on to our next harbour. The morning was calm as we slipped our lines and headed out through the lock gates. However, once we were approaching the mouth of the river, where it flows into the North Sea, a strong wind began to whip. There was a confused slop driven by the strength of the wind and the currents of water flowing from both the Weser and the River Elbe. Beneath us, there were shallow shifting sandbanks. The passage was rough between

Bremerhaven and Cuxhaven, a small harbour at the mouth of the Elbe, and the wind had not been forecast, or else we would not have set sail that day, but *Larry* was built for the weather, her deep, heavy keel keeping her steady in the water.

I had anti-fouled her hull months earlier in Cornwall, to try and keep barnacles and weed from clinging to her planks, creating drag and slowing us down as we sailed. But in spite of my best efforts, I had noticed in the harbour at Bremerhaven that marine life had still found a way to colonise *Larry*. She was coated in barnacles, *Elminius modestus*. Having originated in Australia, these barnacles are an invasive species, first recorded in Europe in 1946 after travelling all the way across the Pacific and the Atlantic on the hull of a ship. Although the barnacles have a larval stage when they drift in the water column, once they settle, they become sessile. Via cement glands in their heads, beneath their shells, they borrow mobility by attaching themselves to another object – to driftwood, a boat, or sometimes to a sea creature – I had seen them encrusting the tails of the humpbacks in the North Atlantic. We were now transporting a small colony of these barnacles, clinging to *Larry*'s hull. They looked almost like tiny flowers, the fixed, ridged plates of their shell light in colour, the operculum at the centre streaked with grey. To feed, they would open their operculum, extending their legs, covered in cilia, out into the water to catch planktonic drifters.

The reduction in speed they caused was palpable and frustrating. As we slammed into wave after wave, I felt as if we might as well be going backwards. I desperately wanted to reach Cuxhaven before night fell, to be able to rest and get a full night's sleep. All the difficulties of sailing through a storm are amplified at night, when you see less and your imagination runs further. With just the two of us on board, we were working hard against the weather. I was at the helm, my left hand clutching the wooden tiller, on which I had inscribed the words 'By the wind, with gentle hands'. But now I was clasping the tiller, my knuckles white, I fought against the weather. My right hand was clamped to the wire of the guard rail, and my legs were braced, the waves rolling and lurching, threatening to dislodge me. Sea spray flew, lashing against me. The summer's day was warm, and thankfully I had made good on my promise to buy new waterproofs, but my hair, face and hands were all soaked.

I cursed the barnacles for making this journey even harder, but a part of me could not help but admire them, clinging resolutely to *Larry*'s hull as we travelled through the thrashing water. In my early years on the Pembrokeshire shores, I had been delighted by the way that they would seal small oceans inside themselves when the tide receded. By locking water inside their shells, they could survive periods of desiccation, before the sea came to wash over them again. Later, at university, I had studied the zonation of the rocky shore, observing the different creatures that thrived at different distances from the sea. Barnacles, mussels and limpets all spent swathes of time underwater, as well as on dry land. They were adapted to be buffeted by the moving water, they thrived in the salt. In rough weather, they stayed steadfast on their rock despite the constant pummelling and pounding of the waves. When the tide was low, they would experience hot sun, wind. They had to build their shells thick to repel the oystercatchers, whose long red beaks would try to infiltrate their safe haven. These were creatures who thrived despite an extreme amount of stress, their environment constantly changing. Resilient. Able to withstand whatever the sea threw at them. They were made for this environment.

And perhaps I was too. My hand hurt where it clasped the wire of the guard rail, my knee was sore where it was braced against the combings of the cockpit. I was tired, hungry, thirsty, as neither of us had been able to take a break for hours. Sometimes we would be hit by a rogue wave that would slam into us from the side, the boat listing, gunnels dipping in the water, my body fighting to keep my place at the helm. And yet I felt safe, capable. I gave up following a compass course. Instead, as Henry kept an eye on the navigation to ensure we were always in safe water and travelling towards our destination, I began to pick my way between the waves, reading the water to give us and *Larry* the smoothest ride. I became transfixed, utterly absorbed in peak and trough, predicting how the water would move, as we moved through it. The world beyond this fell away. It was just me, the boat, the sea. Like the navigation of the seabirds, this experience seemed to bridge some kind of gap in me between innate and learned, drawing on my affinity for the water and the lessons I had spent my life so far learning.

We made it safely into the Elbe River and docked in Cuxhaven, dusk falling as we waited for the wind to blow through. As soon as the weather settled, we slipped our lines and headed upriver to Brunsbüttel. Although the Elbe stretches from the Czech Republic into Germany, connecting Hamburg with the North Sea where we had joined it, Brunsbüttel is the western entrance for the Kiel Canal. The canal is a man-made waterway, completed in 1895. Its function is to allow boats to travel between the North Sea and the Baltic, along a 98-kilometre channel, rather than having to sail 460 kilometres around the northern tip of Denmark. It felt surreal, taking *Larry*, an ocean-going boat, down this tame, tree-lined stretch of water. We had to keep to the bank, allowing great tankers to transit back and forth around us, while hikers poled their way past on the canal path. Private boats are only allowed on the canal during daylight hours, so we reached the eastern end of the canal just in time, right before evening fell. As we circled, waiting for our signal to be able to move forwards and lock out, I felt a wave of anticipation rising. Ahead of me were the large metal gates, the entrance to the lock. When they opened, we passed through, and tied up to a low wooden pontoon. Then they closed behind us, sealing us in. Ahead of us, the next set of gates began to open. The Baltic Sea. A sea that was entirely new to me, and to *Larry*, in all of her years of sailing. It was new to the barnacles too. We had picked up these creatures in salt water, and carried them with us. The Baltic is brackish, far fresher. *E. modestus* have spread so well – from Australia, to the UK, to Holland, to Germany. They are now the dominant species of barnacle in Cuxhaven, where we had made port. However, their foothold disappears at Brunsbüttel, and they still have not managed to infiltrate the bay at Kiel. They cannot survive the low salinity of the water here. By now, they were too far from home, in an environment they were not adapted for. Here we would be able to scrape the hull clean.

As we sailed out into the bay at Kiel, we joined a flurry of sailing boats heading in all different directions. There were traditional boats like *Larry*, fibreglass boats like *Brave* and *Balaena*, steel-hulled vessels like *Song of the Whale*. The sun was beginning to set and the golden hour illuminated the water. I raised my hands in the air in victory as we headed to drop anchor. An old boat in a new sea.

For weeks we moved with the wind wherever we wished, dropping anchor when we found somewhere we wanted to stay, exploring this unfamiliar water. The first time I swam in the Baltic, I was apprehensive about the cold. I hung from the bowsprit, dipping my toes. But it was no colder than the Pembrokeshire water in the summer. I let go of the wooden spar, my body dropping into the cool embrace. I let my head slide under the surface, swimming down. I had my eyes open, and it was strange, not feeling the sting of salt in this brackish water. I swam alongside the hull, running my hand over the coarse barnacles, careful not to let them grate my skin. They were still clinging to the planking, but with a scrub they would come free. I surfaced feeling elated.

Each morning, I started just like this, with a swim before the heat of the day began to build. I walked as far as I could out along the bowsprit, no longer cautious, testing my balance before jumping into the water, washing away my sleepiness. Then I would climb out on to the deck, pulling myself on board by the shrouds, lying on the pine planks seamed with caulking, enjoying the feel of the wood against my back. When I had dried off, I would dress and sit in the cockpit with my morning coffee and some breakfast. Some days, we would hoist the sails, moving *Larry* somewhere new, and others we wouldn't. She flew so well on a light wind, and sometimes we would have friendly races with other sailors on the water, testing her capabilities. Days when we stayed at anchor, I would sit with my notebooks, my paints, to write and sketch. In the afternoons, I would take out our sailing dinghy, the tender to *Larry*, zipping about on the water, hauling up the centreboard that keeps the boat steady to land on beaches.

I was invigorated by this new environment. The abundance of the Pembrokeshire shores is so evident that even for newcomers it is easy to find seals, streams of shearwaters, fire crow on the clifftop fringes. Here, you had to look a little harder, but the life was still there. I would swim over beds of seagrass, the blades waving gently in the water. On closer investigation, it seemed to be the habitat for hundreds of tiny starfish. Fresh-water sponges clung to rocky outcrops, and we frequently saw the rolling fins of harbour porpoise while sailing. There are seals here too, although I have yet to see them: grey seals, harbour seals, and ringed seals, their small bodies marked with patches and rings, hence their name. You can find both fresh-water and salt-water fish in the Baltic: cod and herring, pike and perch. There are no whales resident in these waters, but over the last ten years sightings of humpbacks and fin whales have been increasing. I had heard about Baltic amber, a seam of resin below the Baltic Sea. As the water erodes the sediment covering it, the waves wash pieces of it ashore, earth-toned jewels on the surf. I wanted to see some for myself.

Our world was the boat, the sea around us. Eventually, as summer drew on, we had to head to port. This stretch of time on the water was coming to a close. We sailed further east, making for Rostock. We docked on a sunny August afternoon, the entire waterfront transformed, flags flying everywhere for the Hanse Sail, one of the largest sailing festivals in Europe. We were lucky to find a space, but, once we had tied up, we busied ourselves washing the decks, wiping salt from varnish. We left *Larry* fresh and shining before heading home to work, planning on spending all of our weekends for the rest of the summer and autumn exploring this new stretch of coast on *Larry*, before the cold drove us off the water.

A week after coming off the boat, despite a whole summer when I had felt strong and healthy, in the simple act of turning over in bed at night, I tore the damaged disc in my back. A hot pain ran down my leg, my heart dropping with it, and my lower back immediately began to swell. At first I told myself that my body was lying to me, but, inside, I knew; and an MRI confirmed the problem. Initially I could limp, but once the leaves on the trees began to turn and fall to the ground in autumn, I could barely walk at all. My hopes of being back on *Larry* for weekends evaporated. Once again, my sleep was stolen from me, I could no longer sit, and had to lie down for everything except my meals. Although I had known I would always have to work to maintain the health of my back, I never expected to return to this prone state so soon. This time I had no degree to focus on, no points to prove with my sailing. I cried, worried that, without these things, I would not have the tenacity to get through it again.

Henry went alone to wrap *Larry* up in her winter covers. Her engine was winterised against the Baltic chill, her sails taken off and stowed. I stayed at home, with my physiotherapy and epidurals. My early concerns had been unfounded. Although I no longer had large, ambitious goals to motivate me, I still had the resilience I had been practising for years. This was a setback, but I had staying power. I had been shaped by hardship into somebody more resilient, more confident, with greater respect for my body. Through my experience of sailing, studying, spending time with the whales, time in the water, I had built a shell around myself, and I had attachments that made me feel secure. I could now harbour my own ocean, my own water, to sustain me, until that time when the swells would carry me on my way once again. I saw my back as an obstacle, refusing to let it be anything more than a rock in a stream. It took months of effort before my body started to heal, but it did. As the first snows of winter fell, I knew that, in time, I would be back on my feet.

I wear no watch, but the passage of time is easy to mark, the tide slackening at this southern end of Ramsey Sound. The lowest rocks of the reef across the water are covered once again. This slackening is my favourite. For a scarce few moments, each part of the water seems to be moving a different way, and the possibilities seem endless. Then it stills entirely, before beginning to flood as the tide sets in again. The patterns of the water have now faded to a quiescence. There is nothing left to show they were ever there, save for the knowledge that they will be written each day anew with the falling tide. The sun has dropped, and the shadows begin to grow long, Ramsey Island fast becoming a dark silhouette. The porpoise are close now, feeding up into the tide, their dark backs less than fifty metres away. They start to head to the middle of the Sound, further away than I would expect, but that must be where the fish are.

This place runs like a fault line, right through the centre of me.

Sitting here, it would be easy to imagine that the world is hanging in a perfect balance. That human lives and the marine world exist in a harmony with each other. But perfect moments aren't the whole picture. We have known for too long that our lives cause damage to the oceans. The hourglass of time to change this relationship is nearly empty. The sand now runs faster than ever. Nearly empty, but not yet.

Our oceans, our planet, are facing so many challenges, it is easy and understandable to feel overwhelmed by their weight. In these pages alone, we have explored the changes on my home shores that I have observed in my short lifetime, from the loss of nesting pairs of kittiwakes, to gannets and seals becoming tangled in discarded fishing gear. We have seen the historical scar that whaling has left on the seas, and the current damage we are causing to the whales as the noise generated by humans muffles their song. We have seen how the chemistry of the world's oceans is becoming more acidic, and how by-catch drowns cetaceans, seals and seabirds alike. From my own story, it is easy to see a life that is inextricably linked to the sea, but in reality so are all of our stories. Part of the air we breathe comes from zooplankton in the surface waters, fixing carbon and providing us with oxygen. The oceans are constantly working to cycle carbon so that the atmosphere can support our life. The oceans connect all of the land on our planet. The oceans provide homes for the creatures who fill us with awe and delight.

The scale of the threat is large, but we must not turn away. In informing ourselves of the part we play in the damage, we also inform our choices, and realise that the power we have to do harm is also the power to do good. Dr Sylvia Earle, 'Her Deepness' herself, recently delivered a talk for the Natural History Museum in London. The part that resonated most with me was her profound belief in hope. She has seen more of the seas than most of us ever will, from expeditions at the surface to voyages into the deep ocean. Over her lifetime, she has seen a huge loss in ocean health. In the face of this change, she said, we are given choices. There is a choice to turn towards despair, and assume that things have already gone so far that we as a species are doomed. The planet will go on without us. There will still be sea, however inhospitable we have made it. If we decide that this problem is too big for us, we choose despair. We can choose hope. But this must be an active hope. The worst choice is to hope for change, but to leave it to others, rather than participating in whatever way you can. That choice, that thin edge between hope and despair, is there, every single day.

The choice, the action, looks different for all of us. Dr Earle argues that we must all respond, we must fight for the sea, in the best way we can as individuals. That we should consider ourselves, and look for the area in which we can deliver the strongest contribution. The oceans need artists as much as they need scientists. They need lawyers and politicians as much as they need sailors. They need designers and filmmakers. They need mothers and children. They need shopkeepers and CEOs. They need policymakers in government who will speak for the environment, and mean what they say. They need us all.

For her part, Dr Earle has turned her hope into a physical manifestation through the creation of Hope Spots. Through her organisation Mission Blue, Dr Earle has been working to identify areas of the oceans that are in some way critical for the health of our planet as a whole. This could be because they are important breeding or feeding grounds for specific species. They could be migratory corridors. They could be areas of connectivity, where ocean currents converge to transport nutrients through the oceans, or coral reefs, arcs of life from which abundance flows. Perhaps they contain endemic species, or a huge diversity of life. Once Hope Spots are selected, Mission Blue provides scientific advice and legal advocacy to protect and preserve them, and they are then managed through local organisations.

All around the world, fishermen are making choices to reduce or eliminate by-catch. In 2002, the Southern Seabird Solutions Trust was formed in response to concerns about the destruction of albatross. The trust provides a seat at the table for government, NGOs, seabird scientists, fishermen and fisheries agencies. It works by having fishermen talk to other fishermen, so that they can speak on a level playing field, without feeling patronised or blamed. Through skipper exchanges and workshops, they have developed positive fishing practices that reduce albatross by-catch. There are a variety of methods, which include reducing the amount of bait that the lines are set with in the first place, then making them less attractive to the birds. They also weight the lines, so they set deeper in the water, out of reach of the albatross. They set at night, so that the visual cues for the albatross are not as strong, and they use physical barriers: lines that keep the birds away from the hooks. Many fishermen are open to altering their practices, changing the types of nets and traps they use to minimise by-catch.

For myself, I have had to adapt my choices and my actions

to the way life has taken me and what my body needs. I am no longer on an academic path, but I am sharing my knowledge of the ocean. I am not able to sail as much as I would like at the moment, but I take part in research studies when I can, monitoring the world of the cetaceans, in the hope that by discovering how marine mammals use their habitat, we can develop effective conservation strategies to protect them. In the future, I would love to work in a similar manner to Dr Shane Gero, spending an extended period of time in one area, with one group of whales, getting to know them individually, and then recording and communicating science on all aspects of their biology and ecology.

And part of my contribution lies here, in these words. My aim with this book was to give you an ocean to hold in your hands, no matter how far from the waters you find yourself. The oceans belong to all of us, and so does the care they require. Like the bright-eyed mariner in Coleridge's poem, I hope I have passed on a warning about what we are in danger of destroying. But as well as provoking fear, I hope the book also prompts renewed appreciation for the wonders that the oceans contain, and inspiration that we humans can still change course and steer towards a brighter future.

The sun is falling further, the world descending into shadow with the early winter dark, and I know that soon I shall have to pull myself away from my perch at Pen-Dal-Aderyn, to walk back to the safe harbour of the cottage and company. I am not yet sure what the next part of my own story will be. I will get back to sailing, but I plan to take the steps slowly. In the past, I have looked to the water to provide a passage of healing. I know now that healing does not come through ambitious voyages or grand gestures, but through small, everyday things. It comes through having time to prepare myself nutritious meals, through full nights of sleep. Through gratitude, recognising the good I have around me. I find it in walks through the forest, along the river, the shore. It comes through connection with others and with time for meditation. Healing is in the purr of my cat as she curls close around my back. It drips from the honeysuckle flowers that grow around my first home. It comes with trust in myself, and, as the sun rises each day, it comes with the light bursting through my windows after the hours of dark.

In time, I would love for *Larry* to be a platform to give back to the oceans, to share, to find a way to study from her in a way that supports me, and supports the sea. I am using this healing time to let plans circulate in my head, and to hatch ideas, while also understanding that these schemes may get knocked off course, the tide of my life may turn and send me in a different direction. Wherever I go next, know that I go gently, moving like water.

For the Sea

Individual action is important.

Perhaps it might begin with re-evaluating our relationship with the creatures of the sea, seeing them as wild animals, rather than a plentiful resource to keep taking for our consumption. There are some people on this planet who are entirely dependent on fishing for their protein, but many of us have other choices. I grew up catching fish on the Pembrokeshire coast. At points along my journey, I have had a difficult relationship with food. I am aware that cutting out foods from your diet is a step to take with caution so that it does not cause a spiral of nutrient deficiencies, and that creating rules around food can be a difficult and dangerous thing. But even reducing our consumption of seafood has an impact. In response to diminishing demand, there will be less trawlers, less nets to catch dolphins, porpoise, seals and seabirds. And if you do eat fish, you can make a positive choice to know what you are buying, where it comes from, and how it is caught. You can use your purchasing power to reward more sustainable practices. If we stop eating farmed fish, we reduce the need for swathes of sand eels to be harvested from the ocean, stealing the food of the kittiwakes.

It could be argued that, in doing so, we are destroying the cultural heritage of fishermen. However, the fisheries I am talking about are not small, artisanal businesses, but large-scale, industrial operations. There is no heritage in the sheer amount of fish we currently pull from the oceans. Furthermore, we can look back to the cod fisheries on the Grand Banks, whose mismanagement led to the biggest lay-off in Canadian history, when 30,000 jobs vanished, or to the fishing fleets of Pembrokeshire, which have folded, leaving only small-scale fishing. Unsustainable fishing undermines its own future. It provides jobs for today, without thought of tomorrow.

Reducing plastic consumption and participating in beach clean-ups are simple ways to have an impact, as there is less waste to find its way into the sea. And the less fish we consume, the less we need fishing gear, and the less will be discarded, drowning cetaceans and strangling seabirds. But what about the ghost gear already afloat in our seas? There are charities working around the world to remove it. One of these is a registered charity called Ghost Fishing UK. They have a team of volunteer divers who work to remove lost and discarded fishing gear. This is a very dangerous job, as nets can be huge, and the divers risk being caught and drowned, like the cetaceans and other creatures, so they need to be highly trained and surveys have to be carried out before these recovery operations can begin. On their website, you can both report a net and donate to support their work.

Here are some of the individuals whose actions have most inspired me:

Dr Sylvia Earle: her Hope Spot project aims to identify and protect areas of marine diversity. Anyone can nominate a Hope Spot and anyone can donate money to support them. Any donations received are spread through all of these Hope Spots.

Dr Laura Feyrer and Dr Shane Gero are using their science to help understand the oceans, to improve our understanding, to inform policy, and perhaps even to be able to communicate with whales.

Rachel Carson was a woman whose writing inspired many to turn towards the sea. Her seminal work *Silent Spring* is an incredibly powerful piece of literature that sparked an entire environmental movement and the banning of DDT.

Justine Willeford, a woman who I connected with over a shared love of the sea, despite there being a whole ocean between us, makes her contribution through her work as the founder of Pelican House. She designs and produces ethically made swimwear, designed by women, for women. Her company provides apparel through which we can access and play in the sea, and also gives back. Every purchase provides a donation to her conservation partner, Island Conservation, who work to remove invasive species from islands and prevent extinction.

Arianna Liconti is a passionate marine ecologist who has volunteered for various marine NGOs while studying for her Bachelors and Masters degrees. I recently spoke with her as she headed home through the hills of Liguria after a busy day working for the sea. Although she does produce academic publications, she has really found her strength in connecting people.

She believes, as I do, that nature, that the sea, does not exist in a vacuum, and that its health is our health. She is now Head of Science and Ecosystem Manager at OutBe. OutBe is a new start-up, making their contribution through business. They work for people and planet, to link researchers to outdoor enthusiasts. They encourage people to get into the outdoors, on both land and sea, for their own health. This time spent outdoors is then used to collect data which is sent back to researchers. For example, they work with The Ocean Race, formerly the Whitbread Round the World Race. The participating boats are fitted with sensors that record data on sea surface temperature and salinity. This information is then fed back to scientists, who as a result don't have to conduct a field study of their own. If a scientist is researching the growth of kelp but cannot get into the field every day, OutBe can link them with an outdoor centre that leads snorkelling trips in the study area, and the snorkellers participate in the science and collect the data for them.

Emilie Ehrhardt is a new graduate in Wildlife Filmmaking. She is using her talents to tell stories for the planet. Not every solution for the sea starts with the sea: what we do on land is equally significant, as it all feeds back into the global climate. In her debut short film, titled *Urban Eden*, Ehrhardt showcases an example of how nature can thrive in an urban environment. In her childhood home, the city of Copenhagen, she uses her film to present CopenHill. The hill is a man-made mountain, with a green wild slope, and a power plant inside. The plant takes waste from the city, which is then burned to produce energy. The steam that is made during this process is then filtered to such a high grade that when it is released over the hill, it is in fact cleaner than the city air. They are working to develop a carbon-capture system, which will sequester the carbon produced in this process, and also in the surrounding area. This man-made hill is a hope spot of its own. The soil covering the hill is 95 per cent building waste, and was originally planted with sixty species of plants native to Denmark, modelled on the ecological make-up of a wild area nearby. Within just one year, nature took hold and the number of species doubled. Ehrhardt's film shows that when nature is included in the design of our spaces, it can thrive. CopenHill is a wild oasis in the cityscape, and makes nature accessible both on foot and by wheelchair. Ehrhardt's film is an inspiration to consider what nature-based solutions we can start in our own gardens and communities, wherever we live.

And finally, my own mother, Jackie Morris, throughout her career, has used her paintings to celebrate the natural world. She has produced work for Greenpeace, and, at the same time as my back was crumbling and she was driving me home from surgery and caring for my bedridden self, she was illustrating a book written by Robert Macfarlane called *The Lost Words*. She describes the book as a beautiful protest against the removal of nature words from the *Oxford Junior Dictionary*. Her contribution was a huge catalyst in crowdfunding movements to buy a copy of this beautiful book for every primary school in the UK, in order to connect children to nature. The book has gone on to inspire not just children, but a BBC Proms performance and two folk albums by the Spell Songs collective, as well as countless individuals.

This is a network, mostly of women, that I have built up over the years. I find their contributions an inspiration for positive change. But if this short list illustrates anything, it is that there is no single way to work for the sea. Every individual effort counts. Every individual effort is a gesture of hope. And, taken together, all our individual efforts can make a difference and bring about positive change.

Bibliography

1 Fire Crow

Berta, A., 2015, *Whales, Dolphins and Porpoise: A Natural History and Species Guide*, Sussex, UK, Ivy Press

Boness, D. J., James, H., 1979, 'Reproductive behaviour of the grey seal (*Halichoerus grypus*) on Sable Island, Nova Scotia', *Journal of Zoology*, 188:477–500

Frederiksen, M., Wanless, S., Harris, M. P., Rothery, P., Wilson, L. J., 2004, 'The role of industrial fisheries and oceanographic change in the decline of North Sea black-legged kittiwakes', *Journal of Applied Ecology*, 41:1129–39

Griffith, S., 1990, *A History of Quakers in Pembrokeshire*, Dyfed, UK, Gomer Press

Heubeck, M. A., Mellor, R. M., Harvey, P. V., Mainwood, A. R., Riddington, R.,1999, 'Estimating the population size and rate of decline of kittiwakes, *Rissa tridactyla*, breeding in Shetland', *Bird Study*, 46:48–61

Howell, D. W., 2019, *An Historical Atlas of Pembrokeshire: Volume 5*, Haverfordwest, UK, Pembrokeshire County History Trust

Johnstone, I., Mucklow, C., Cross, T., Lock, L., Carter, I., 2011, 'The return of the Red-billed Chough to Cornwall:

the first ten years and prospects for the future', *British Birds*, 104:416

RSPB, 'Cornish Choughs', https://www.rspb.org.uk/birds-and-wildlife/wildlife-guides/bird-a-z/chough/cornish-choughs/, accessed 27/05/21

Rus Hoelzel, A., 2002, *Marine Mammal Biology: An Evolutionary Approach*, Oxford, UK, Blackwell Publishing

Sandvik, H., Reiertsen, T. K., Erikstad, K. E., Anker-Nilssen, T., Barrett, R. T., Lorentsen, S., H., Systad, G., H., Myksvoll, M. S., 2014, 'The decline of Norwegian kittiwake populations: modelling the role of ocean warming', *Climate Research*, 60:91–102

Whale and Dolphin Conservation, 'By-catch', https://uk.whales.org/our–4-goals/prevent-deaths-in-nets/goodbye-by-catch-what-you-need-to-know/, accessed 30/5/21

Notes and Resources

For more information on Ramsey Island and its choughs: https://www.rspb.org.uk/reserves-and-events/reserves-a-z/ramsey-island/

2 Sperm Whale

Berta, A., 2015, *Whales, Dolphins and Porpoise: A Natural History and Species Guide*, Sussex, UK, Ivy Press

Carson, R., 1950, *The Sea Around Us*, USA, Oxford University Press, 2018

Carson, R., 1962, *Silent Spring*, Boston, Houghton Mifflin, 2002

Carson, R., 1941, *Under the Sea Wind*, London, Penguin Classics, 2007

Davies, R. W., Rangeley, R., 2010, 'Banking on cod: exploring economic incentives for recovering Grand Banks and North Sea cod fisheries', *Marine Policy*, 34:92–8

Dwyer, A., 2012, 'Atlantic borderland: natives, fishers, planters and merchants in Notre Dame Bay, 1713–1802', doctoral dissertation, Memorial University of Newfoundland

Estes, J. A., Demaster, D. P., Doak, D. F., Williams, T. M., Brownell, R. L., 2006, *Whales, Whaling, and Ocean Ecosystems*, California, University of California Press

Feyrer, L. J., 2021, 'Northern bottlenose whales in Canada: the story of exploitation, conservation and recovery', doctoral thesis, Dalhousie University

Feyrer, L. J., Zhao, S. T., Whitehead, H., Matthews, C. J., 2020, 'Prolonged maternal investment in northern bottlenose whales alters our understanding of beaked whale reproductive life history', *PLOS One*, 15:e0235114

Hersh, T. A., 2021, 'Dialects over space and time: cultural identity and evolution in sperm whale codas', doctoral thesis, Dalhousie University

Hooker, S. K., Fahlman, A., Moore, M. J., Aguilar De Soto, N., Bernaldo de Quirós, Y., Brubakk, A. O., Costa, D. P., Costidis, A. M., Dennison, S., Falke, K. J., Fernandez, A., 2012, 'Deadly diving? Physiological and behavioural management of decompression stress in diving mammals', *Proceedings of the Royal Society Biological Sciences*, 279:1041–50

Rus Hoelzel, A., 2002, *Marine Mammal Biology: An Evolutionary Approach*, Oxford, UK, Blackwell Publishing

Panneton, W. M., 2013, 'The mammalian diving response: an enigmatic reflex to preserve life?', *Physiology*, 28:284–97

Myers, R. A., Hutchings, J. A., Barrowman, N. J., 1997, 'Why

do fish stocks collapse? The example of cod in Atlantic Canada', *Ecological Applications*, 7:91–106
Nakashima, B. S., Wheeler, J. P., 2002, 'Capelin (*Mallotus villosus*) spawning behaviour in Newfoundland waters – the interaction between beach and demersal spawning, *ICES Journal of Marine Science*, 59:909–16
Pinet, P. R., 2013, *Invitation to Oceanography*, sixth edition, Burlington, Massachusetts, Jones and Barrett Learning
Smith, G., 1983, 'The International Whaling Commission: an analysis of the past and reflections on the future', *Natural Resources Law*, 16:543
Watwood, S. L., Miller, P. J., Johnson, M., Madsen, P. T., Tyack, P. L., 2006, 'Deep-diving foraging behaviour of sperm whales (*Physeter macrocephalus*)', *Journal of Animal Ecology*, 75:814–25

Resources

Project Ceti, https://www.projectceti.org/
Mission Blue, documentary directed by Robert Nixon and Fisher Stevens
Mission Blue: The Sylvia Earle Alliance, https://mission-blue.org/

3 Human

Berta, A., 2015, *Whales, Dolphins and Porpoise: A Natural History and Species Guide*, Sussex, UK, Ivy Press
Rus Hoelzel, A., 2002, *Marine Mammal Biology: An Evolutionary Approach*, Oxford, UK, Blackwell Publishing
Sini, M. I., Canning, S. J., Stockin, K. A., Pierce, G. J., 2005, 'Bottlenose dolphins around Aberdeen Harbour, north-east Scotland: a short study of habitat utilization

and the potential effects of boat traffic', *Journal of the Marine Biological Association of the United Kingdom*, 85:1547–54

Stockin, K. A., Weir, C. R., Pierce, G. J., 2006, 'Examining the importance of Aberdeenshire (UK) coastal waters for North Sea bottlenose dolphins (*Tursiops truncatus*)', *Journal of the Marine Biological Association of the United Kingdom*, 86:201–7

4 Wandering Albatross

ACAP, 2012, 'Wandering albatross', http://acap.aq/en/acap-species/304-wandering-albatross/file, accessed 10/01/22

Berrow, S. D., Huin, N., Humpidge. R., Murray, A. W., Prince, P. A., 1999, 'Wing and primary growth of the Wandering Albatross', *The Condor*, 101:360–8

Cherel, Y., Xavier, J. C., de Grissac, S., Trouvé, C., Weimerskirch, H., 2017, 'Feeding ecology, isotopic niche, and ingestion of fishery-related items of the wandering albatross, *Diomedea exulans*, at Kerguelen and Crozet Islands', *Marine Ecology Progress Series*, 565:197–215

Coleridge, S. T., 1798, *The Rime of the Ancient Mariner*, New York, USA, Dover Publications, 1992

Croxall, J. P., Rothery, P., Pickering, S. P., Prince, P. A., 1990, 'Reproductive performance, recruitment and survival of wandering albatrosses, *Diomedea exulans*, at Bird Island, South Georgia', *Journal of Animal Ecology*, 775–96

Croxall J. P., Prince P. A., 1990, 'Recoveries of wandering albatrosses, *Diomedea exulans*, ringed at South Georgia 1958–1986', *Ringing & Migration*, 11:43–51

De Roi, T., Fitter J., Jones, M., 2008, *Albatross: Their World, Their Ways*, Cardiff, UK, Firefly Press

Frankish, C., 2021, 'Movement ecology and fisheries by-catch risk of albatross and large petrel species from Bird Island, South Georgia', doctoral dissertation, University of Cambridge

Froy, H., Lewis, S., Catry, P., Bishop, C. M., Forster, I. P., Fukuda, A., Higuchi, H., Phalan, B., Xavier, J. C., Nussey, D. H., Phillips, R. A., 2015, 'Age-related variation in foraging behaviour in the wandering albatross at South Georgia: no evidence for senescence', *PLOS One*, 10.1:e0116415

Jones, M. G., Dilley, B. J., Hagens, Q. A., Louw, H., Mertz, E. M., Visser, P., Ryan, P. G., 2017, 'Wandering albatross *Diomedea exulans* breeding phenology at Marion Island', *Polar Biology*, 40:1139–48

Pickering, S. P., Berrow, S. D., 2001, 'Courtship behaviour of the wandering albatross, *Diomedea exulans*, at Bird Island, South Georgia', *Marine Ornithology*, 29:29–37

Jones, M. G., Dilley, B. J., Hagens, Q. A., Louw, H., Mertz, E. M., Visser, P., Ryan, P. G., 2014, 'The effect of parental age, experience and historical reproductive success on wandering albatross (*Diomedea exulans*) chick growth and survival, *Polar Biology*, 37:1633–44

Nevitt, G. A., Losekoot, M., Weimerskirch, H., 2008, 'Evidence for olfactory search in wandering albatross, *Diomedea exulans*', *Proceedings of the National Academy of Sciences*, 105(12):4576–81

Pickering, S. P., Berrow, S. D., 2001, 'Courtship behaviour of the wandering albatross, *Diomedea exulans*, at Bird Island, South Georgia', *Marine Ornithology*, 29:29–37

Prince, P. A., Wood, A. G., Barton, T., Croxall, J. P., 1992, 'Satellite tracking of wandering albatrosses (*Diomedea exulans*) in the South Atlantic', *Antarctic Science*, 4:31–6

Prince, P. A., Weimerskirch, H., Huin, N., Rodwell, S., 1997, 'Molt, maturation of plumage and ageing in the Wandering Albatross', *The Condor*, 99:58–72

Rackete, C., Poncet, S., Good, S. D., Phillips, R. A., Passfield, K., Trathan, P., 2021, 'Variation among colonies in breeding success and population trajectories of wandering albatrosses, *Diomedea exulans*, at South Georgia', *Polar Biology*, 44:221–7

Richardson, P. L., Wakefield, E. D., Phillips, R. A., 2018, 'Flight speed and performance of the wandering albatross with respect to wind', *Movement Ecology*, 6:1–5

Stone, D. W., Gunn, C., Nord, A., Phillips, R. A., McCafferty, D. J., 2021, 'Plumage development and environmental factors influence surface temperature gradients and heat loss in wandering albatross chicks', *Journal of Thermal Biology*, 97:102777

Thomson, C., 2019, *Sea People: In Search of the Ancient Navigators of the Pacific*, London, UK, William Collins

Weimerskirch, H., Barbraud, C., Lys, P., 2000, 'Sex differences in parental investment and chick growth in wandering albatrosses: fitness consequences', *Ecology*, 81:309–18

Weimerskirch, H., Åkesson, S., Pinaud, D., 2006, 'Postnatal dispersal of wandering albatrosses *Diomedea exulans*: implications for the conservation of the species', *Journal of Avian Biology*, 23–8

Weimerskirch, H., Cherel, Y., Delord, K., Jaeger, A., Patrick, S. C., Riotte-Lambert, L., 2014, 'Lifetime foraging patterns of the wandering albatross: life on the move!' *Journal of Experimental Marine Biology and Ecology*, 450:68–78

5 Humpback Whale

Baker, C. S., Flórez-González, L., Abernethy, B., Rosenbaum, H. C., Slade, R. W., Capella, J., Bannister, J. L., 1998, 'Mitochondrial DNA variation and maternal gene flow among humpback whales of the Southern Hemisphere', *Marine Mammal Science*, 14:721–37

Berta, A., 2015, *Whales, Dolphins and Porpoise: A Natural History and Species Guide*, Sussex, UK, Ivy Press

Brewer, P. G., Hester, K., 2009, 'Ocean acidification and the increasing transparency of the ocean to low-frequency sound', *Oceanography*, 22:86–93

Carson, R., 1950, *The Sea Around Us*, USA, Oxford University Press, 2018

Clapham, P. J., 1996, 'The social and reproductive biology of humpback whales: an ecological perspective', *Mammal Review*, 26:27–49

Falkowski, P., Scholes, R. J., Boyle, E. E., Canadell, J., Canfield, D., Elser, J., Gruber, N., Hibbard, K., Högberg, P., Linder, S., Mackenzie, F. T., 2000, 'The global carbon cycle: a test of our knowledge of earth as a system', *Science*, 290:291–6

Forward, R. B., 1988, 'Diel vertical migration: zooplankton photobiology and behaviour', *Oceanography: Marine Biology Annual Review*, 26:1–393

Fournet, M. E., Matthews, L. P., Gabriele, C. M., Haver, S., Mellinger, D. K., Klinck, H., 2018, 'Humpback whales *Megaptera novaeangliae* alter calling behavior in response to natural sounds and vessel noise', *Marine Ecology Progress Series*, 607:251–68

Gazioğlu, C., Müftüoğlu, A. E., Demir, V., Aksu, A., Okutan, V., 2015, 'Connection between ocean acidification and sound propagation', *International Journal of Environment and Geoinformatics*, 2:16–26

Heenehan, H., Stanistreet, J. E., Corkeron, P. J., Bouveret, L., Chalifour, J., Davis, G. E., Henriquez, A., Kiszka, J. J., Kline, L., Reed, C., Shamir-Reynoso, O., 2019, 'Caribbean Sea soundscapes: monitoring humpback whales, biological sounds, geological events, and anthropogenic impacts of vessel noise', *Frontiers in Marine Science*, 347

Helweg, D., Jenkins, P., Cato, D., McCauley, R., Garrigue, C., 1998, 'Geographic variation in South Pacific humpback whale songs', *Behaviour*, 135:1–27

Ilyina, T., Zeebe, R., Brewer, P., 2009, 'Changes in underwater sound propagation caused by ocean acidification', IOP Conference Series, *Earth and Environmental Science*, Vol. 6, No. 46, IOP Publishing

Medwin, H., Clay, C. S., Stanton, T. K., 1999, 'Fundamentals of acoustical oceanography', *Journal of the Acoustical Society of America*, 105, 2065–6

Munk, W., Wunsch, C., 1979, 'Ocean acoustic tomography: a scheme for large-scale monitoring', *Deep Sea Research*, Part A, Oceanographic Research Papers, 26:123–61

CDC, 2019, 'Noise and hearing loss prevention', https://www.cdc.gov/nceh/hearing_loss/what_noises_cause_hearing_loss.html. accessed 1/3/22

Parsons, E. C., Wright, A. J., Gore, M. A., 2008, 'The nature of humpback whale (*Megaptera novaeangliae*) song', *Journal of Marine Animals and Their Ecology*, 1:22–31

Rossi, T., Connell, S. D., Nagelkerken, I., 2016, 'Silent oceans: ocean acidification impoverishes natural soundscapes by altering sound production of the world's noisiest marine invertebrate', *Proceedings of the Royal Society Biological Sciences*, 283:2015.3046

Rus Hoelzel, A., 2002, *Marine Mammal Biology: An Evolutionary Approach*, Oxford, UK, Blackwell Publishing

Stevens, A., 2014, 'A photo-ID study of the Risso's dolphin (*Grampus griseus*) in Welsh coastal waters and the use of Maxent modelling to examine the environmental determinants of spatial and temporal distribution in the Irish Sea', MSc thesis, Bangor University

Strutt, J. W., Baron Rayleigh, 1896, *The Theory of Sound*, London, UK, Macmillan

Thomas, P. O., Reeves, R. R., Brownell Jr, R. L., 2016, 'Status of the world's baleen whales', *Marine Mammal Science*, 32:682–734

Resources

Songs of the Humpback Whale, a music album of whale song, released by Roger Payne, 1970

More information about *Song of the Whale* and Marine Conservation Research and their projects can be found at http://www.marineconservationresearch.co.uk

The Dominica Sperm Whale Project, http://www.thespermwhaleproject.org/

6 Shearwater

Foster, J. J., Smolka, J., Nilsson, D. E., Dacke, M., 2018, 'How animals follow the stars', *Proceedings of the Royal Society Biological Sciences*, 285, 1871:2017.2322

Grant, R. G., 2018, *Sentinels of the Sea: A Miscellany of Lighthouses Past*, London, UK, Thames and Hudson

Matthews, G. V., 1964, 'Individual experience as a factor in the navigation of Manx shearwaters', *The Auk*, 81:132–46

Popova, M., 2019, *Figuring*, New York, USA, Pantheon Books

Sauer, E. F., 1958, 'Celestial navigation by birds', *Scientific American*, 199:42–7

Thomson, C., 2019, *Sea People: In Search of the Ancient Navigators of the Pacific*, London, UK, William Collins

7 Barnacle

Barnes, H., Barnes, M., 1960, 'Recent spread and present distribution of the barnacle *Elminius modestus* Darwin in north-west Europe', *Proceedings of the Zoological Society of London*, Vol. 135, No. 1, 137–45

Den Hartog, C., 1953, 'Immigration, dissemination and ecology of *Elminius modestus* Darwin in the North Sea, especially along the Dutch coast', *Beaufortia*, 1:4

De Roi, T., Fitter J., Jones, M., 2008, *Albatross: Their World, Their Ways*, Cardiff, UK, Firefly Press

Foster, B. A., 1971, 'Desiccation as a factor in the intertidal zonation of barnacles', *Marine Biology*, 8:1

For the Sea

Resources

Mission Blue: The Sylvia Earle Alliance, https://mission-blue.org/

Laura Feyrer, https://feyrer.weebly.com/

Shane Gero, http://www.shanegero.com/

Pelican House, https://www.pelicanhousesc.com/

Arianna Liconti and OutBe, @ari_liconti https://www.outbe.earth/

Emilie Ehrhardt and Urban Eden, @wild.and.about @urban_eden_film

Jackie Morris, https://www.jackiemorris.co.uk/

Acknowledgements

Firstly, my most sincere thank you to Jessica Woollard. You saw this story as the notes and drawings in my field journals, and encouraged me to work from there towards the book we have today. Without you, *Move Like Water* would never have made it out into the world. Thank you for your trust and encouragement, from the very first time I walked into your office, my head spinning, having barely stepped ashore on return from sea. To Nicola Davies, for always encouraging me, for always supporting my ideas and putting those that seemed unreachable within arm's length. Laura Barber, your insight while editing this book is beyond anything I could have ever hoped for. Thank you for taking a chance, and letting me take you to sea through these pages. The story has always felt in the safest hands, which has allowed me to explore so much more than I ever thought I would.

My thanks to Ffion Rees, for my first boat job, and the continued support you give. To Gabriel Clarke, for the opportunity to sail across the North Sea, and the help you gave me when I was buying *Brave*. I will never forget how you told me that you knew I knew more than I thought I did. Although our paths diverged, we both made our lives on the sea, exactly how we said we would. To my Ancient Mariner, although you will never read these pages, the impact you had on my life was profound, and I will never sail the North Sea without thinking of you. To have shared your last voyage with you was an honour. Thank you to Dominic and Barbara Bridgman, for the early lessons. I hope I will join you on *Lyhner* one day. For Professor Hal Whitehead, Dr Laura Feyrer, Dr Mauricio Cantor and Verity Thomson, the crew with whom I sailed on board *Balaena*. Thanks for the outstanding time with the whales. This trip was such a pivotal point in my life and a catalyst for so much more. I feel so lucky to have sailed aboard *Balaena*. Thank you to *Song of the Whale*, Marine Conservation Research, Richard McLanaghan, Anna Moscrop, Oliver Boisseau, Niall MacAllister and Judith Matz. I hope I have done *Song of the Whale* justice, to acknowledge the pioneering work that you have done for the sea for almost fifty years. I will never forget what a privilege it is to have been a small part of that legacy. To Dr Clare Embling, for advising me at university. Dr Shane Gero, thank you for always giving me your time so generously, and for the wonderfully creative, open-minded work that you do. Thank you for encouraging me to develop my field journals. To Dr Taylor Hersh, for producing your beautiful thesis, and the time you took to talk to me about sperm whales.

Thank you to Justine Willeford, Arianna Liconti, Emilie Ehrhardt and Alys Perry. I will never not find you inspiring, and am in awe of how you wonderful women use your talents to work for the sea. Emilie, Alys, and Alicia Leaman, thank you for the check-ins. Writing this book has taken me to some deep places, and although they are worth exploring, that hasn't been without its challenges. Thank you for the dinner calls, the drinks, the supportive texts, the care packages. Although we are far-flung, you have never felt far away. Thank you to Dawn Brown and Isobel Pullin. I hope this book makes you proud.

To my family. To my mother, Jackie Morris; I think these

pages already explained what for. To my father, Tim Stowe. Thank you for moving to Ramsey: if you hadn't made that choice, maybe I would never have grown up in Pembrokeshire at all. Thank you for the books, and the poetry you gave me. I have always savoured the words. To my grandmother; I wear your locket every day.

To Sue, and the Singer family, for taking care of *Larry* so well, and for letting us be the next part of her story.

And lastly, to Henry Carey-Morgan. I'm sure it has something to do with magpies. Thank you for letting me be a part of your story, as you are of mine. Thank you for believing in this book, before it was a book, and the support you have given me through the process. Every day is an adventure.